MISSIONS
REMEMBERED

OTHER McGRAW-HILL MILITARY HISTORY/AVIATION TITLES

Aviation Week & Space Technology
MILITARY AIRCRAFT PILOT REPORTS

Edwards Park
ANGELS TWENTY: A YOUNG AMERICAN FLYER
A LONG WAY FROM HOME

Jerry W. Cook
ONCE A FIGHTER PILOT . . .

John L. Stewart
THE FORBIDDEN DIARY: A B-24 NAVIGATOR REMEMBERS

Raymond F. Toliver and Trevor J. Constable
THE BLOND KNIGHT OF GERMANY

Edward C. Herlik
SEPARATED BY WAR: AN ORAL HISTORY BY DESERT STORM FLIERS
AND THEIR FAMILIES

Larry Reithmaier
MACH I AND BEYOND: THE ILLUSTRATED GUIDE TO
HIGH-SPEED FLIGHT

Leroy Cook
AMERICAN AVIATION: AN ILLUSTRATED HISTORY, SECOND EDITION

MISSIONS REMEMBERED

RECOLLECTIONS OF THE WORLD WAR II AIR WAR

THE MEN OF THE MIDDLE TENNESSEE

WWII FIGHTER PILOTS ASSOCIATION

McGraw-Hill
New York   San Francisco   Washington, D.C.   Auckland   Bogotá
Caracas   Lisbon   London   Madrid   Mexico City   Milan
Montreal   New Delhi   San Juan   Singapore
Sydney   Tokyo   Toronto

Library of Congress Cataloging-in-Publication Data

Missions remembered : recollections of the World War II air war / by
 the men of the Middle Tennessee WWII Fighter Pilots Association.
     p.   cm.
  ISBN 0-07-001649-6
  1. World War, 1939–1945—Aerial operations, American. 2. Fighter
pilots—United States—Tennessee—Biography. 3. World War,
1939–1945—Personal narratives, American. I. Middle Tennessee WWII
Fighter Pilots Association.
D790.M556 1998
940.54'4373—dc21   97-48517
                                                CIP

McGraw-Hill

A Division of The **McGraw·Hill** Companies

1 2 3 4 5 6 7 8 9 0 DOC/DOC 9 0 3 2 1 0 9 8

ISBN 0-07-001649-6

*The sponsoring editor for this book was Shelley Carr, the editing
supervisor was Jane Palmieri, and the production supervisor was Sherri
Souffrance. It was set in Fairfield by Dennis Smith of McGraw-Hill's
Professional Book Group composition unit, Hightstown, NJ.*

Printed and bound by R. R. Donnelley & Sons Company.

Contents

Preface

It has indeed been an honor to be involved with the preparation of this book and to work with the individuals whose memories and accomplishments are collected here. It is important that their deeds be recorded for posterity and that we remember and honor these men who served so gallantly more than a half century ago. It is very easy for those of us who were not involved to forget the courage and sacrifices made by these boys who were forced to become men virtually overnight. It is particularly important as the events of World War II fade into the carpet of time and the generation of warriors who performed these deeds passes on. I hope that this work will help us keep the memories alive and prevent such a catastrophe from ever taking place again.

I would like to express my personal gratitude to Mr. Stephen Gossett and Ms. Phyl Taylor for their invaluable help in assembling, proofreading, and typing much of this material. Without their assistance, and the cooperation of the faculty, staff, and students in the Department of Aerospace at Middle Tennessee State University, this book would not have been possible.

Ronald J. Ferrara, Ed.D.
Department of Aerospace
Middle Tennessee State University
Murfreesboro, Tennessee

Introduction

Early in 1992, a small group of Nashvillians, all of whom had been fighter pilots during World War II, decided to establish an informal organization of their peers. Thus came into being what is known as The Middle Tennessee WWII Fighter Pilots Association, and the ranks have now swelled to approximately 30 members.

Without intending any exclusivity whatsoever, there are three screens or requirements for membership:

1. fighter pilots only
2. World War II participation
3. actual combat experience

A strong case can be made that World War II was the most significant event of the twentieth century. Everything that happened prior to it was prelude, and everything since then has been affected by it. Think about it.

These fellows formed this group primarily for their own purposes: recollection of similar and shared experiences, nostalgia, fellowship, and fun. It is, however, highly appropriate that those of us who were not privileged to have shared their experiences firsthand have an opportunity to become familiar with them. These are some of their stories. Read them with excitement and, above all, with pride. It's a huge honor for me to be associated with these men.

The following quotation is a brief excerpt from "Locksley Hall" by the celebrated English poet Lord Alfred Tennyson.

For I dipt into the future, far as human eye could see,
Saw the Vision of the world, and all the wonder that would be:
Saw the heavens fill with commerce, argosies of magic sails,
Pilots of the purple twilight, dropping down with costly bales;
Heard the heaven fill with shouting, and there rain'd a ghastly
  dew
From the nations' airy navies grappling in the central blue;
Far along the world-wide whisper of the south-wind rushing
  warm,
With the standards of the peoples plunging thro' the thunder-
  storm;
Till the war-drum throbb'd no longer, and the battle-flags were
  furl'd
In the Parliament of man, the Federation of the world.

[1835–1842; published 1842]

Tennyson penned these words approximately 70 years before the Wright brothers' first flight, 80 years before World War I, and 100 years prior to World War II. And does the last line quoted here underscore his clairvoyance about the establishment of the United Nations?

In any case, perhaps his words will provide a fitting introduction to the experiences these men have relived more than 50 years later.

John K. Breast, Honorary Member
The Middle Tennessee WWII
Fighter Pilots Association

These Damn Yanks

Told by Enoch B. Stephenson, Jr.
Major, U.S.A.A.F.

During the late 1930s, there was a great deal written about Germany's preparations for war. Adolf Hitler had assumed complete dictatorial control of Germany and had invaded and taken control of Austria. Everyone assumed he would invade his neighbors because he spoke on numerous occasions of the need for more space for German people. There were many, though, who seemed unable to grasp the idea that World War II was imminent. During the summer of 1939, my good friend Bedford Lochridge and I were working for Mr. Burke Wilkes, who was a Maury County surveyor. On the morning of September 1, I heard on the radio that the Germans had invaded Poland, and the word *blitzkrieg* became a part of the languages. En route to our job site, we had a long and serious talk about the implications for us of the event; after all, both Bedford and I were vulnerable. In the car with Mr. Wilkes, we had an extended discussion of the situation and its implications for a couple of 21-year-old guys. The consensus was that America would be actively involved in the war in the future, but we had no idea when.

I believe my interest in flying began when Charles Lindbergh flew his famous flight from New York to Paris non-stop in May 1927. I recall my mother telling me all that was in the newspapers and showing me the pictures. I was 9 years old at the time and was in a state of awe. Lindbergh later wrote a book entitled *We*, in which he told the complete history of the preparation for the flight and detailed the actual flight. My mother read me everything about Lindbergh that was printed.

In the fall of 1927, Lindbergh flew the *Spirit of St. Louis* over the entire United States, passing over our school. Our teachers took us out onto the playground so we could see the plane fly overhead. That is one of those occasions that made a lasting impression on me, and I still remember it to this day. Many years later, when I was in the Tennessee Air Guard, Ernest Perry and I flew Major General Frank McCoy of Nashville to Andrews Air Force Base. We delivered a portrait of Lieutenant General Frank Maxwell Andrews, a native of Nashville who was killed in the crash of a B-24 Liberator in Iceland en route from London to Washington. The portrait hangs in the officers' club at Andrews.

During the course of lunch at the club, Frank McCoy spotted General George Kinney, for whom Frank had worked as an intelligence officer in the Pacific theater during World War II, having lunch with Charles Lindbergh. Frank asked if Ernest and I would like to meet them, and of course we said yes. We interrupted them long enough to shake hands, which was indeed a great treat.

During my youth, I was fascinated with the exploits of Jimmy Doolittle, who was doing such great things to advance the future of aviation. Little did I realize that I would fly P-51s under his command in the Eighth Air Force. He was probably the greatest pilot of all time. Certainly, he was in my opinion.

During my high school days, two friends who were a bit older showed up in the pinks and greens of the uniforms of the Army Air Corps wearing silver wings. I was hooked. I determined that I would get at least 2 years of college so I could become an aviation cadet. An ancillary benefit was that it would enable me to avoid being an infantryman.

After an inauspicious career in college, I went to work for Glenn L. Martin Company in Maryland and gathered my necessary paperwork to join the Naval Air Corps. Things went smoothly until I went to Anacostia Naval Air Station, District of Columbia, to take a physical, and the doctor turned me down because of low blood pressure. I immediately applied to the Army Air Corps and used the knowledge that had been imparted to me by the Navy flight surgeon to pass the Army physical. This was the fall of 1941, and I was impatient for the call to duty. Then December 7 occurred when the Japanese attacked Pearl Harbor and I felt that it would not be long. Sure enough, right after the first of the year, I was instructed to report for induction on January 10, but was delayed for 2 days by some administrative matter.

My swearing in on January 12 was inauspicious, but it meant I was one step on the way to becoming an Army Air Corps pilot. Next morning, the four of us who had met at the recruiting office boarded the train for Chicago and points west. Our trip to Chicago overnight was uneventful.

At Chicago, we changed trains for the long part of the journey to Williams Army Air Field, Chandler, Arizona. None of us had any idea where Chandler was situated or even where the state of Arizona was. Our journey took us through Illinois, Missouri, and Arkansas to Texarkana. Then we traveled west through more of Texas than I thought existed to El Paso. Incidentally, the distance from Texarkana to El Paso is farther than from Chicago to Texarkana. From El Paso we went west across New Mexico and southern Arizona, finally arriving at Chandler. After 5 days and nights of no showers, we were glad to get an opportunity to shower and spruce up.

Our activity at Williams was called *preflight*, which consisted of early rising, calisthenics, close order drill, and more of the same. Fortunately, preflight for us lasted only 2 weeks; I heard of many guys who spent from 6 weeks to 6 months in preflight.

About the first of February, we boarded the train for our primary flight school at Ontario, California. The school was named Cal Aero and was a civilian flight school under contract to the Air Corps. All flight and ground school instruction was conducted by civilians, and administration and discipline were in the hands of Air Corps officers and noncoms.

Classroom instruction covered engines, aerodynamics, meteorology, navigation, flight rules, and Army regulations. One week we would have ground school in the morning and flying in the afternoon and then the reverse the next week. Ground school was a complete bore. Flying was another matter. The first time I had ever been off the ground was my orientation flight. It was with the civilian flight commander, a really fine man. Later, I heard he had said something to the effect that he knew I would make it because I couldn't get the grin off my face for the duration of the 1-hour flight. It was a great thrill.

Primary flight school was in the Stearman PT-17, an open-cockpit biplane that required helmet and goggles. It was a rugged bird. It had to be to withstand the rough landings and aerobatic maneuvers through which we put it. It was a good trainer because it responded well to the controls. It took a lot of punishment and it did great aerobatics. I enjoyed primary thoroughly, despite the constant fear I would do or fail to do something that would wash me out of the program. It was very disheartening to see guys missing from formations because they could not cut the mustard. Basic flight training was conducted at the same location. We flew the Vultee BT-13, a low-wing plane with a closed canopy, wing flaps, a 450-hp engine, and a propeller that operated in two positions which was controlled from the cockpit. It was good to move forward.

By the time we had 4 months of training, we had flown 120 hours in two different airplanes. We had learned takeoffs, climbs, traffic patterns, approaches, and landings. We had learned to stall the airplane and to recover, to spin the airplane and to recover. Additionally, we were taught simulated forced landings and the rudiments of bailing out and using the parachute. The really fun part was learning aerobatics, which consisted of loops, chandelles, Immelmanns, lazy eights, pylon eights, slow rolls, barrel rolls, snap rolls, and other maneuvers that we all found useful in combat.

In early June, we were transferred to Luke Army Air Field, Phoenix, Arizona. It was another trip to the desert, but this time in the heat of summer. There we flew the North American AT-6, with a 650-hp engine equipped with a constant speed propeller which enabled us to select whatever engine speed was indicated for the flight condition. In addition to the maneuvers learned in primary and basic, we learned basic instrument flying and formation flying in two-ship elements and in flights of four ships. Luke was back to the Army; we were in GI barracks and ate in a GI mess hall. We marched from one formation to another, rolled out at reveille, saluted everything that moved, and did our best to adapt to Army life.

Finally, after 6.5 months of rigorous training and the constantly nagging fear of washing out came graduation day. What a thrill it was to get our wings and commissions. I was on top of the world and had 30 days of leave at home. But about a week or so into that leave, I was bored to tears. Of course, I was glad to be with the family and all of that, but we had run out of things to talk about and I was ready to move on. Finally, about September 1, I reported at Myrtle Beach Army Air Field, South Carolina, to join the 20th Fighter Group. There we checked out in the Curtiss P-40, an obsolescent fighter which was made famous by General Claire Chenault and the Flying Tigers in China. After about a month, we were transferred to Pinellas Army Air Field, St. Petersburg, Florida, to fly the same airplanes. We were redesignated the 337th Fighter Group and were assigned the mission of instructing a class of new pilots every 2 months in fighter tactics so they could be prepared for combat. I was fortunate in being retained as an instructor and thus was able to build up my experience in flying, instruction, and leadership.

About September 1, 1943, four of us were assigned to the 339th Fighter Group in Walterboro, South Carolina, to train in the Aircobra, the Bell P-39, which we flew to the desert at Rice, California. Rice was a railroad stop halfway between Needles and Blythe, which is to say in the middle of nowhere. But the weather was always clear, we could fly all the time, and we were able to get some superb training. After about 8 months on the desert, we boarded another troop train for the port of embarkation, New York City. All of us thought we would be assigned to the Ninth Air Force to fly close support for the ground forces because that was the type of training we had conducted. However, after we boarded H.M.S. *Sterling Castle* and were out, the group commander told us that we were assigned to the Eighth Air Force and would fly North American P-51 Mustangs. We really got a kick out of that bit of information. I had a little experience in the Mustang while flying accelerated service testing of the airplane at Patterson Field, Dayton, Ohio, in the fall of 1942.

The trip across the North Atlantic was uneventful and very boring. It took about 10 days and nights in convoy. We arrived at Glasgow in mid-March in typical, unpleasant British weather, boarded a train, and arrived at our lovely village of Fowlmere, Herefordshire, in more rather inclement weather. Fowlmere RAF Station had been an airfield during World War I and had reverted to farming between wars to be reclaimed by the RAF during the Battle of Britain. The RAF had turned the base over to the Eighth Fighter Command for our use, and we had been preceded by service troops who prepared it for us.

Immediately, Mustangs began to arrive, and we began our transition and familiarization flights. I was commander of A Flight and supervised the qualification of the members of the flight who had been with us for the last year and the replacement pilots who were assigned. Our first operational mission was flown on April 30, 1944, and was an orientation sweep just into the coastal areas of France and the Lowlands. However, it was not long before we were flying real tough missions into France and Germany. On May 21, my close friend, the squadron operations officer, was shot down because he broke the cardinal rule that said not to go back for a second pass. As a result of his bailing out, I became squadron operations officer. Our group commander had successfully gotten all of the pilots a promotion as soon as we arrived in England. As soon as I was made squadron ops, he put me in for promotion to major. Fighter command made him wait 3 months and resubmit the recommendation, and it came through post haste.

The P-51 was the finest fighter plane of World War II. It had enough fuel to escort the bombers to any of the targets they were assigned. It had firepower: six .50-caliber machine guns capable of destroying any target except the German Tiger tanks. It had maneuverability sufficient to dogfight with any German fighter plane until the ME-262 jet came along. It had a marvelous engine, the Rolls Royce Merlin, which was built in the United States by the Packard Motor Car Company. It had great dependability and reliability. All of the pilots loved flying the Mustang.

During the summer months, the British observed double British summer time, which was 1 hour earlier than daylight-saving time. Considering that London is on about the same parallel of latitude with Hudson Bay in Canada, we had lots of daylight. It was not uncommon for us to be alerted at 0330 for an 0530 takeoff. We took comfort in the fact that the poor devils flying the Fortresses and Liberators had been up much longer. Generally, our missions consisted of our group of three squadrons, composed of 16 aircraft each. Lead of the group was rotated, and usually the group commander, Colonel John B. Henry, would lead the group or would designate one of the three squadron commanders or squadron operations officers to serve as group lead.

During the summer of 1944, the Germans were feeling the pinch of lack of fuel for their air force, but that did not mean they didn't frequently mount forceful attacks on our bombers. On the contrary, it was not uncommon on a raid on Berlin to encounter large and formidable formations of ME-109s and FW-190s. These were the first line fighters of the German Air Force and were excellent aircraft, but they lacked the necessary attributes to best the Mustang. Of course, much depended on the skill of the pilot. Many of the German fighter pilots had experience dating back to the Spanish Civil War in the mid-1930s, and they were very skillful and were able to inflict heavy damage on the bombers when there were no Allied fighters to drive them off.

Deflection gunnery, or shooting at the enemy fighter from an angle, was taught stateside before going overseas. However, when we got into combat, we learned rather quickly that the surefire way to be successful was to shoot from directly behind the enemy. If we could get close enough behind the enemy, we could hit him in the wing root area, underneath the fuselage where the fuel pump was located. A hit there from the six .50-caliber machine guns of the Mustang would result in fire, or at least total engine failure as a result of fuel starvation. Needless to say, this approach is easier to describe and write than it was to execute in the heat of battle.

While the German attacks were mounted from the side of the bomber formation, they later adopted the head-on approach. They would get far enough off the bomber formation to enable themselves to approach the leader of the bombers in a slight dive and inflict heavy damage. This tactic resulted in the development of the chin turret and nose turret for both the B-17s and B-24s. How successful these developments were I do not know, but I am sure they were helpful from a psychological perspective.

During May 1944, my squadron, the 513th Fighter Squadron of the 339th Fighter Group, flew 30 missions over the Continent, 25 of which were bomber escort missions. My logbook indicates that I flew 13 of these missions. The longest was 5 hours 35 minutes to Berlin and return to escort bombers. Flak (antiaircraft fire) was extremely heavy in all of Germany, but was especially so around Berlin. The German 88-mm gun was the best gun in the war and was used as a field artillery piece and as a tank gun in addition to its use as an antiaircraft gun. It was very effective when coordinated with effective radar.

During June, the squadron flew 41 missions, mostly bomber escort over targets in France. The purpose of most of these missions was to interrupt and disrupt rail and road transportation. Of course, D-Day, June 6, 1944, was the most important event of the war. It was a 6.5-hour mission, taking off before dawn to patrol the western sector of the invasion area. We flew at 1500 to 2000 ft to stay below the cloud deck. The weather was lousy and it took a gutsy decision by General Eisenhower to say "Go" instead of delaying another day. But for us, it was a very uneventful mission. I felt very secure in my Mustang foxhole in the sky. After some 3 hours of patrolling, we were relieved by the other two squadrons of our group, whereupon we proceeded to the area of Normandy, south of the invasion area, to search for targets of opportunity. After cruising around at 1500 ft for a while, I spotted a unit of Germans in the standard six-man vehicle of the Wehrmacht. They were en route to the beachhead at a rapid clip, but we overtook them handily, set up a traffic pattern, and rendered them hors de combat. We had been ordered to remain well away from the beach and the invasion area, so we patrolled for a while and returned to base by swinging far to the west. Back at Fowlmere, we found a flight order waiting for us calling for another mission of the same format.

We were unfortunate on another mission to have two of our pilots lose radio contact and overfly England. We were sure that they finally ditched in the Atlantic Ocean, west of the British Isles.

During the summer, after the threat of German fighters had subsided, we were ordered to leave the bombers after they had dropped their bombs and turned for home. Our mission then became what was called "Chattanooga Choo-Choo." It was so labeled because we were directed to destroy as much rolling stock as we could find. Of course, at that time, all locomotives were powered by steam and they would blow sky high when we hit them. The Germans, in an effort to thwart us, would string cables on either side of the tracks and then hang loose cables to the ground. This meant we had to break off our strafing runs sooner in order not to fly through the cables. One of our pilots, Lieutenant Paul Fiorito, did in fact fly through and wrap one of the dangling cables around his propeller spinner. He pulled up and noticed a horrible vibration, but found that the engine was still running, so he retarded the power and reduced the engine revolutions per minute. He realized that he was still flying, so he decided not to jump until the engine quit completely. It didn't quit and he didn't jump, so he wound up landing at the nearest airfield in Britain, a Royal Air Force station. The British received him with hospitality, and after they had an opportunity to examine the plane, they were shaking their heads saying "these damn Yanks."

At times, the weather in England would sock in so badly that missions were scrubbed. I remember one morning when I was group leader, I received a phone call from the leader of another group at another base. He was concerned about the forecast for takeoff, climb, and return. So was I. After a bit, he brought me up short by saying if the bombers were going, we had a duty to go with them and lend our support. We went, had a successful mission, and a safe and uneventful return.

The summer and fall were routine. Enemy activity fell off, except for the flak. It was always fierce, accurate, and deadly. We were able to find a few enemy airfields with parked aircraft, which we strafed, but we couldn't stay long because of the flak. The Germans were so short of aviation fuel that we would shoot up their planes, but they would not burn. We believe we did considerable damage, but we didn't get credit for aircraft destroyed. Our only appreciable leisure time was 3 or 4 days to go to London. It was and is an incredible city. There were so many historic things to see that I was always amazed. I did get a week of leave in Edinburgh, and it was fine except that I was homesick for my friends back at the base, but I stuck with it because the flight surgeon thought I needed it.

I finished my combat tour on October 17 and returned to the States shortly thereafter. I was sent to the training command, but it was closing down rapidly. It was quite a letdown after combat flying and having so many close friends and lots of activity. Once again, I found I was homesick for my friends.

Last Cadet
to Solo

Told by Charles R. Mott
Captain, U.S.A.A.F.

Class 42-G was put together hurriedly in December 1941 from the new cadets stationed at Maxwell Army Airfield, Montgomery, Alabama. I joined as a cadet on December 26, 1941. Just prior to New Year's Day, approximately 200 new, young, eager flying cadets departed for Dorr Field, Arcadia, Florida, which had just been completed for the Embry-Riddle Aeronautical Institute, a private company with a contract to train cadets through the primary phase of pilot training.

Soon after arrival, each of us was assigned to a flight instructor. My instructor was a man named Richard E. Life. Mr. Life was a very good pilot with a sense of compassion for cadet life during pilot training. I will never forget my first flight with Mr. Life and all the exciting times, such as when he briefly let me take the controls of the Stearman PT-17 that we were flying at the time.

Before long, Mr. Life began training all five of the cadets assigned to him. We practiced landings, takeoffs, turns, flight patterns, and the most exciting part of all, loops, chandelles, snap rolls, slow rolls, the entire menu of aerobatics. In my evaluation, he believed I was doing everything fairly well except for landing the airplane. Until you can land the airplane, you cannot solo. Slowly, I watched the other four cadets solo, and believe me, I began to get concerned. I was making good progress in all of the instructions I received except that every time I landed the Stearman, I landed on the left wheel and then bounced to the right wheel and bounced back and forth down the runway.

Finally, the day of reckoning came and I had completed approximately 13 hours of flight training. The class generally had soloed after 10 or 11 hours of training. Some of them soloed with as few as 8. As I reported to the flight line on January 19, 1942, Mr. Life told me he was putting me up for a "check ride." He explained to me that I had learned everything with the exception of a smooth landing. He further explained, "I have tried every way in the world to detect what you are doing wrong in handling the stick and rudders during touch-down." I thought he was just being gentle and kind and letting me down easy, for it was by tradition that a check ride was a formality, preceding a cadet being "washed out" of pilot training. I explained this concern to Mr. Life and he assured me this was not the case, since it was his hope that the Army pilot who gave me the check ride would be able to detect what I was doing wrong during the landing maneuver.

After Mr. Life introduced me to the Army pilot, I climbed into the cockpit. We took off and the lieutenant said, "Let's go to auxiliary field No. 2 and shoot some landings." This was encouraging to me because he was going right to the problem that Mr. Life described and was not asking me to demonstrate all the training I had received up to that time. And now the landing! Sure enough, I did it again, landing on the left wheel, bouncing to the right wheel, and bouncing left and right down the runway. Only a Stearman PT-17 could survive this kind of abuse during a landing.

We taxied around to the end of the runway, and the check pilot said, "On this next landing, I want you to sit square in the center of the seat and sit at attention. Move only your eyes left and right during the landing procedure." I have to admit this was the first time I had ever been ordered to "sit" at attention except during the meal hour. I said, "Yes, sir. I understand." I took off and flew the landing pattern sitting at attention on the final approach, moving only my eyes left and right. I set the airplane down for a perfect three-point landing. The instructor said, "That was fine. Let's do it again." I landed a second time with a perfect three-point landing. The instructor told me to taxi to the parking area and stop. He said, "I think we have found the problem. You have been leaning out to the left because you're so damn short you can't see the ground. Instead of relying on all of your senses while doing this, you were unconsciously taking the stick to the left, slightly dropping the left wing, and landing on the left wheel." He said further, "I believe you can fly this thing by yourself." I couldn't believe his words. I was going to solo!

He asked me, "If you undershoot the landing, what are you going to do?" and I replied, "Give her the gun." "What if you find yourself in the middle of a landing and you don't know what the hell to do?" I answered, "Give her the gun." He patted me on the back, stepped down from the wing, and walked away. I taxied to the takeoff position, brimming with enthusiasm, and gave her the gun. Making my first solo takeoff and flying around the flight pattern was an experience words cannot describe. I whistled and sang the whole time. If I landed the airplane successfully, I would be the happiest guy in the world. I turned final approach sitting at attention, moving my eyes from left to right, as I let the Stearman descend to another perfect three-point landing. I taxied over, picked up the lieutenant, and flew back to the main field at the Dorr Field complex. As we taxied in, the lieutenant congratulated me and told me that he would tell Mr. Life what the problem was.

I never saw Mr. Richard E. Life again and I have long since forgotten the name of the Army pilot lieutenant who gave me the chance to solo. I went on to basic and advanced flying schools and received my commission as a second lieutenant, U.S. Army Air Corps on August 5, 1942, with the classification of 1055 (fighter pilot).

Iirmgard

Told by Lee V. Gossick
Captain, U.S.A.A.F.

During the summer of 1942 until the American invasion in western North Africa in November, there were two American fighter groups supporting General Montgomery's Eighth Army in the effort to defeat the German and Italian forces led by General Field Marshall Rommel in their objective of capturing the Suez Canal and the oil resources of the Middle East. The first American fighter group to join the action was the 57th. Along with a number of fellow graduates of the class of 42-D of the Army Air Corps Aviation Cadet program, I joined the 57th in New England in May of 1942 where they had been preparing for deployment. Our aircraft was the Curtiss Wright P-40E, which was the only Army fighter in production in any quantity at that time. In June, the 57th was sent to Egypt, and the new "sprogs," as we were called by the old-timers, were assigned to the newly formed 79th Fighter Group. In September, the 79th followed the 57th to the Egyptian desert just west of Alexandria. Our P-40s joined those being flown by units from nearly all of the countries of the British Commonwealth, Australia, New Zealand, South Africa, and Canada. The LTK units were equipped with Hawker Hurricanes and that wonderful bird, the Spitfire.

The primary mission of the P-40 units was ground attack and bomber escort. The fighter opposition was mainly the

ME-109, with the less frequently encountered FW-190 and Italian Macchi 200. The ME-109 was always far above us attacking with the sun behind and with airspeed and climbing capability that gave it a great advantage. The P-40 was a dog at anything above 12,000 or 13,000 ft. It wasn't much of a weight saving, but we removed the oxygen systems as unnecessary baggage.

However, the P-40 had its virtues. It was built like a bridge, and you didn't have to worry about it coming apart no matter how many Gs you pulled. It would easily turn inside the ME-109 if you could get him to engage in a dogfight. You could stay with him if he chose to dive out of an engagement, but he could easily break off by climbing away. In the firepower department, I don't believe any of us would have traded our six .50-caliber guns for his .20-mm cannon.

My unit, the 87th Squadron of the 79th Group, had the good fortune of being able to put back into flyable condition an ME-109G. It had been belly-landed on a salt flat in southern Tunisia. There was no evidence of battle damage; there was still fuel in the tanks, and when put back on its landing gear, the engine easily started, notwithstanding a very badly bent propeller. Maybe the pilot had just had enough of the way things were coming apart for the Germans in North Africa! The name *Iirmgard* was on the side of the aircraft and was probably the name of the pilot's wife or girlfriend. Our squadron mechanics and other specialists worked many long extra hours in making the necessary repairs, scrounging needed parts from other damaged ME-109s left behind by the Luftwaffe. Their work was truly outstanding and included the very necessary preparation of procedures and the checklist needed for evaluating our pilots.

As this work was proceeding, we devised a plan that would hopefully result in every pilot of our squadron flying *Iirmgard* in mock dogfights against a P-40. Captain Fred Jory, one of our senior flight leaders, was tasked to prepare a list of our pilots which ranked, in descending measure, his degree of confidence that the pilot would not screw up and cause us to lose this very useful training resource. It is a testament to Captain Jory's experience and skill in evaluating people and their strengths and weaknesses that the program was completed without incident—until the very last guy on the list was scheduled. The operation of the ME-109s landing gear was as simple as pushing a red button to retract and a green button to lower the gear. Alas, our hapless pilot took off, pushed the green button, and then started doing slow rolls at low altitude. He did not respond to frantic calls on the radio. He finally decided to land after having pushed the red button and, ignoring red flare pistol warnings as he approached on final, submitted *Iirmgard* to another belly landing.

Despite the known history of two "sudden stoppages" of the big fuel-injected Jumo engine, *Iirmgard* was again repaired by our sister squadron, the 86th, and used in similar training with their pilots. That program was cut short, however, by orders from Ninth Air Force Headquarters in Cairo to release *Iirmgard* for delivery to Wright Field at Dayton, Ohio, for testing by the structures laboratory. Drop tests were conducted to evaluate the landing gear and related structure. Tests were run to the point of destruction of the tail section. It is my understanding that it was found that the design criteria applied by the German engineers were more conservative than those practiced by American designers. This finding was applied to fixes found necessary to correct structural problems encountered in the tail section of the P-47 during its early operations in Europe.

Browsing through my pilot's logbook, I find three flights recorded 56 years ago during which I became acquainted with *Iirmgard* (as the squadron operations officer, I could see when *Iirmgard* was available but the next scheduled pilot was not). During my last flight, accompanied by a P-40, we spotted a C-47 on its way to deliver mail and supplies to the forward units. We staged a brief dogfight in front of the "gooney bird" and its crew, which probably made their day and gave them something to write home about!

At the end of the North African campaign, it is recorded that the air-air kill ratio was 9-1 in favor of the Allied fighter units. We suffered significant losses from ground defenses during strafing and dive-bombing operations, but I firmly believe that our 79th Fighter Group's air-to-air record was enhanced by the contribution to our training that *Iirmgard* made possible.

How It All Began
OR Don't Call Us, We'll Call You

Told by Hensley Williams
Major, U.S. Marine Corps

From my teenage years, I was always interested in airplanes and flying. I would often ride my bicycle out to the old McConnell Field (now McCabe Municipal Golf Course) and volunteer to sweep and clean up the hangar, wash down airplanes, and so forth. In exchange, they would occasionally give me a free airplane flight of a few minutes duration in one of the locally owned or controlled aircraft such as a Waco, a Travelair, or even a Ford Trimotor. What excitement! I was there one day when Lindbergh came through and a few times when Army Air Force pursuit or bomber airplanes landed and stayed a day or two. What ecstasy! I soon began to dream of a career in military aviation.

Later, while attending Vanderbilt Engineering School (2 years) and the U.S. Naval Academy (graduated February 7, 1941), I "bummed" airplane rides whenever I could. Following graduation from the U.S.N.A. (10 months to the day before Pearl Harbor), we were required to train as ground Marines for 2 years before being permitted to attend flight training. I was assigned to the field artillery and was transferred from base to base for the next 18 months, for various training, before being sent to the southwest Pacific in August 1942. At the time of each transfer, I always notified Marine Corps headquarters that I desired to go to flight training. I always received a short reply stating my request had been received and would be acted upon when and if aviators were needed and that I need not apply again.

I reached the Guadalcanal combat area of the southwest Pacific in September 1942 and, with heavy heart, had almost given up my dreams of being a combat fighter pilot. However, as the old saying goes, sometimes "the Lord moves in strange ways, His wonders to perform." Soon, the combat losses of the Marine fighter pilots on Guadalcanal became large and Marine Corps headquarters did notify me to take a flight physical immediately and, if passed, to go to flight training. Needless to say, I was ecstatic.

The rest is pretty much history. In about 18 months, I was back in the mid-Pacific flying combat missions, some 52 in all by war's end, as a Marine fighter squadron commander (VMF-113).

The moral of this story is that I owe a great deal to "The Good Lord" and to "Lady Luck"; not only in World War II, but also in Korea and Vietnam.

That First Mission

Told by Edward F. Jones
Captain, U.S.A.A.F.

It was a long, long way from Nashville, Tennessee, to the briefing room on the airstrip at Marina di Grosseto, Italy, as I suited up for my first combat mission as a pilot in the 64th Fighter Squadron, 571st Group, 12th Air Force. The journey started at Miami Beach for Army basic training; a stint in college training detachment at the University of Tampa; classification and preflight school at San Antonio; primary flying school at Ft. Stockton, Texas; basic and advanced flying at Enid and Altus, Oklahoma; combat assignment staging at Lincoln, Nebraska; P-47 school at Wendover, Utah; overseas staging at Baton Rouge, Louisiana; departure at Newport News, Virginia, for an 18-day convoy junket to Oran, North Africa; transfer to a British troop-carrying ship for a 3-day cruise to Naples, Italy; and four nights in a tent city until a C-47 troop carrier provided transportation to Grosseto. The accommodations were spectacular.

Grosseto, on the white and sandy beaches of the Italian Riviera, had been a summer resort for upscale Italians. We pilots were quartered in what had been summer vacation villas, and the ground crews and enlisted men lived in a nearby dormitory of what had been a boarding school for boys. We bunked up two pilots to a room with such creature comforts as hot and cold running water, electricity (provided by our own generator), and an arrangement with remaining natives to keep the villas cleaned and the laundry done. The centerpiece of the complex was a large beachside villa which we appropriated for an officers' club, bar, and dining room. After hearing of combat conditions where crews lived in tents, huts, and worse, I felt that I had been blessed indeed with such a billet.

For a few days after arrival, the other replacement pilots and I spent time in orientation sessions learning about the country and the war in the Mediterranean. We had lectures on the formations and tactics used by the 64th and received some flying time to get the feel of a P-47 with a full combat load.

Then one evening, as happened every evening, the intelligence office posted the names of the pilots for missions the next day. There I was, one of eight pilots on the first flight out. Early the next morning, a six-by-six truck carried seven other pilots and me to the airstrip. The briefing room was in the remains of a hangar that had been bombed and was dominated by a large map on one wall. The briefing officer gave us our targets: two bridges and a rail junction in the Brenner Pass at the foothills of the Alps. The Brenner Pass was the main supply line between Germany and the German troops in Italy, and one of our missions was to keep the flow of supplies disrupted.

Looking at the big map to plot the course from Grosseto to the Brenner Pass, my eye kept going back to the substantial number of red splotches all over northern Italy. These were known concentrations of antiaircraft placements (flak spots) to be avoided if possible. Of course, the Germans were constantly relocating this artillery, so there were always a few surprises in store.

Reality set in when the briefing officer showed me a cabinet with a pigeonhole that had my name on it and told me to leave my billfold and all personal things in the slot until I returned from the mission. In exchange, I was given an escape kit, a flat canvas pocket-sized package that contained, among other things, German and Italian currency, a compass, map, rubber coated hacksaw blade that could be hidden anally, and a fake picture of me in civilian clothes that could be used to create a bogus ID card.

We got our formation positions. I was the No. 4 plane in the second of two 4-ship sections, a position commonly known as Tail End Charlie. We picked up our parachutes and helmets at a trailer just outside the briefing room, and a couple of Jeeps gave us a ride to our respective airplanes. We taxied out to the end of the runway single file, did the preflight check and the engine run-up, and everyone signaled ready for departure. When my turn came and I got out on the end of the runway, I stood up on the brakes, opened up the big 2000-hp engine until the plane was dancing on the ground, released the brakes and applied all the power left, and after what seemed an eternity, the controls began to feel light and I knew I was off the ground. I was the last plane to take my place in the formation, and we were off to the Brenner Pass.

Once we crossed the Apennines Mountains, the bomb line where the Germans on the north and the Allies on the south were deadlocked in a prolonged ground struggle, we turned on our gun switches, armed our bombs, and those of us flying wing for the more experienced pilots began a curious little weaving motion to sweep the skies for signs of enemy aircraft.

The plane felt heavy. I was not used to flying a plane with a 500-lb bomb under each wing, six rockets, a 110-gal belly tank, and the wings full of .50-caliber ammunition for the eight machine guns. On the way to the target, we were shot at twice by 88-mm antiaircraft, but no one was hit, and while the exploding shells got my attention, I felt no panic.

Approaching the target, a rail line snaking through the valleys of towering mountains on either side, the mission leader signaled us to get in single file and prepare for the dive run. The first plane peeled off at a steep angle followed by the rest of us trying to spot the bridges we were to take out. The first plane in the string released its bombs and started to pull out of the dive. Unexpectedly, the bombs detonated a concealed German ammunition dump of massive proportions. From my lofty position at the end of the string, I saw a sudden enormous explosion, far greater than could be caused by any bombs we were carrying, boil up from the valley and engulf the first three planes in our string of eight.

The section leader radioed for the rest of us to jettison our bombs and form up on him for a return to Grosseto. The trip back to home base was uneventful, but I kept seeing in my mind those three out-of-control aircraft tumbling earthward and missing from our return flight. We were routinely debriefed and the location of the tragic event recorded. After V-E Day, the wreckage of the aircraft and the graves of the pilots were located from our description of the site. For a few weeks, the tale of this flight was told over and over, but after I had 10 or 15 missions under my belt, I finally began to believe that this was an especially tough mission and that we were not going to lose three of the eight planes every time we flew.

North Africa

Told by Lee V. Gossick
Captain, U.S.A.A.F.

On April 29, 1942, the graduation exercise at Foster Field, Victoria, Texas, sorted out those who would go to fighters and those who would not. Somehow, I never doubted that I would be among those chosen for fighter assignments. At 0800 hours, our class of 42-D was commissioned and given our wings. We also received our orders for our next assignments.

At 1000 hours, the line formed at the base chapel for the weddings that were to occur. Ruth and I were among them. Each ceremony was allowed 15 minutes. We were instructed to clear the base by 1800 hours because the new class of incoming cadets was short of housing.

My orders were to report to the 57th Fighter Group, Logan Airport, Boston, Massachusetts. Along with other new pilots, we checked out in the P-40 and trained with the "old sports" of the 57th Group until it was deployed to Egypt in July 1942. The 57th was assigned to support the British Eighth Army in its confrontation with Rommel's Afrika Corps, which was threatening the Suez Canal. The 79th was formed and manned with the new pilots and ground crew personnel that were left behind by the 57th. In September 1942, the 79th was ordered to join the action in Egypt. The advance party of pilots arrived only a few days before General Montgomery's all-out counterattack began in the desert on October 23.

In our cots at Landing Ground 174, some 30 mi from Alexandria, the opening artillery barrage from the 882 big guns of the British Eighth Army jolted us out of our sleep. In awe, we watched the night sky illuminated by the barrage. We realized that each of us would soon be a part of the struggle that had been under way for years in the North African desert.

The battle of El Alamein in Egypt was a turning point in the war. We were indeed fortunate to be a part of the "chasers" rather than the "chased." The highlight of my experience during this time involved a combination of events: the birth of our son and the shootout of all shootouts during my combat tour.

Letters from Ruth indicated that our first born was expected about June 6, 1943. On June 10, I received a Marconigram announcing the birth, on June 2, of our son, Roger V. Gossick. It was late in the afternoon and we were in the process of celebrating this event in the officers' club tent when our 87th Squadron was ordered to launch a fighter sweep over the area between Cape Bon, Tunisia, and the island of Pantelleria.

The ensuing action is best described by an excerpt from *A Hostile Sky,* a book by Don Woerpel on the history of the 79th Fighter Group in World War II. While there are many other war stories that come to mind, this one—starting that afternoon of June 10, celebrating the birth of my son—holds a special niche in my memory.

The group had been in the air almost continually for better than 14 hours, and this was their first real action in an otherwise uneventful (for them at least) early June day. Five minutes later, the Skeeters would electrify the entire Northwest African Air Force when they reduced the enemy's fighter strength by 15 in as many minutes.

Sixteen Skeeter Warhawks, their pilots answering today to the call sign "clatter," had droned out of dusty El Houaria earlier in the afternoon to begin their last sweep of the day. Sixteen young men with hard-boned faces burned bronze by the hot desert sun; wearing yellow life jackets, faded khaki shirts, and leather helmets; pistols holstered around their shoulders, hunting knives strapped to their legs, and scarves of discarded parachute silk around their necks.

They were now 5000 ft over the northeast corner of smoke-shrouded Pantelleria when a lone MC-202 (probably a straggler from the fight the Skulls had run into earlier) came stooging through the Skeeters' top section. Paul McArthur pulled around on it and fired several bursts. Pieces flew off the Macchi and coolant poured out. The Alabama pilot gave it another squirt. The Italian rolled the plane over on its back and bailed out, trailing a partially opened chute. One down.

A few minutes later, the Argus eyes of Lieutenant Colonel Charles Grogan sighted a large hospital plane hugging the waves 20 mi off the coast. Hovering around it were 10 ME-109s. Grogan jinked his P-40 once, then again to clear his tail, and tore down after the Messerschmitts. The old Warhawk, considered by many as obsolete even before Pearl Harbor, had its faults but diving with a full head of steam wasn't one of them. With Grogan closing fast, a German went into aerial convulsions trying to evade him, turned left when he should have turned right, and the P-40's wing tip sliced neatly through the 109's control surfaces. Grogan headed for home missing several feet of his port wing while the German cartwheeled into the sea. Two down.

Then suddenly, the entire sky became a jumble of twisting, turning, diving, climbing aircraft. Morris Watkins, flying on Grogan's wing, rode the tail of a 109, his guns knocking large chunks of metal fabric off the German fighter, and the 109 went in beneath a plume of water and debris. That made three! The six wing guns of Johnny Krisch tore another 109 to wreckage and the enemy fighter went into a death plunge. Four down! Kensley Miller got the fifth, a 109, and Asa Adair chased another to within 20 mi of Sicily before the Messerschmitt went into the water with a dead pilot at the controls. Number six! Frank Huff had his yellow section (Huff, Charlie Jaslow, Leo Berinati, and John Dzamba) at 7000 ft when he spotted two 109s orbiting over a downed pilot and reduced their number by one. Seven! Leo Berinati was within gliding distance of Sicily when he caught the other with three long bursts, and the 109 half rolled into the sea. Eight...and the fight was just heating up.

Porky Anderson got number nine, a Macchi that George Lee and Ed Fitzgerald had been harassing, then kicked hard right rudder and went head on for another. This Italian would leave his bones on the floor of the strait as the Macchi came apart under the savage pounding of the Warhawk's six .50s. Ten...and Anderson wasn't through yet. He caught a 109 low on the water and moved in close, his tracers lacing out in red threads to walk a stitched pattern up the side of the Messerschmitt. A short puff of dirty black smoke gasped from the belly of the 109 followed by a tongue of flame. Then there was a burst of bright orange, and the Gustav (an ME-109G) came apart in the air. The squadron's score for the day now stood at eleven.

With the sky full of parachutes, Francis Hennin and I found eight 109s milling around another hospital plane. Hennin's guns stubbornly refused to fire at a crucial moment (he finally got them working by vigorously rapping on the switch panel, but by then, his target was out of range). I exploded one of the German fighters for number twelve and damaged another that hurriedly left the scene pouring coolant.

As abruptly as it started, the fight was over. One moment the sky had been a wild place echoing to the sound of chattering guns and the high-pitched whine of over 30 straining engines. Now it was as silent as a cathedral at midnight.

Paul McArthur had already gotten a Macchi early in the fight. During the battle, he got another and added two 109s and damaged still one more before he was forced to bail out of his shot-up P-40. That brought the score to fifteen destroyed and two damaged with McArthur the only Skeeter pilot down. He had bailed out of his damaged X-83 some 30 mi off the coast of Tunisia, shucked off his chute harness, and clambered aboard a half-inflated dinghy. Overhead orbited me and Adair until relieved by Anderson and Berinati. Jesse Jory and I soon returned in two freshly refueled aircraft to resume the vigil.

The dusk was already deepening into a murky darkness when the air-sea rescue lads sat down beside McArthur in their sputtering Walrus. But his troubles weren't over yet. The Walrus pilot tried every trick in the book in a futile attempt to get off the water, but the ancient flying boat obstinately refused to become airborne. McArthur helped man the plane's bilge pumps as they started taxiing to shore. Then the engine quit. In company with his would-be rescuers, Paul McArthur spent an uncomfortable night bobbing about in the Mediterranean. Early the next morning, a British destroyer tossed them a line and they were towed into port.

Goofs and Foolishness (How Did I Ever Survive?)

Told by Johnnie B. Corbitt
Captain, U.S.A.A.F.

M y first flight ever was with my primary flight instructor, Mr. Floyd Fahey, on Friday, November 13, 1942, six days after my 20th birthday, at Chickasha, Oklahoma. You would think I would be permanently jinxed with the first day of flight instruction on Friday the 13th, but with wonderful instructors and a great deal of good fortune, in spite of my apparent efforts to do myself in, I was able to live my fantasy of becoming a pilot in the Army Air Corps. The following are a few of the incidents that come to mind as I think back over that period of my life on the road to becoming a combat fighter pilot.

Shortly after I began to fly solo, I read the manual on how to do inverted flight. I decided that it would greatly impress my instructor if I practiced the maneuver and surprised him when he started to teach me how to fly upside down. I had learned that to pull back on the stick would cause the nose of the plane to come up in front of me toward the sky and to push it forward caused it to go away from me toward the ground. I studied all of the moves to roll the plane over, and the next time up, I tried them out. They worked beautifully, but I forgot one thing. When the nose started down, naturally,

I pulled back on the stick. The nose came up in front of me all right but headed for the ground. I hung on for dear life and completed a split-S. I waited for instruction from then on regardless of how insignificant the maneuver.

During primary flying school, we had three check rides, two with civilian check pilots and one big one with an army check pilot. Everyone said, "Don't get Lieutenant Cope. He's rough." Naturally, I drew Lieutenant Cope. My instructor told me to stay away from the hills to the north of the field and to go south and west to flat farmland for the simulated forced landing that was sure to come. Everything went well except that Lieutenant Cope directed me north over the hills in spite of everything I could do. Suddenly, the throttle was jerked back and the plane lost power. Try as I might, I couldn't budge the throttle and we were rapidly losing altitude. Lieutenant Cope yelled into the gosport, our method of communication (one way, of course), "Forced landing." I practically passed out but started looking for a suitable field on which to land the plane. There weren't any. I finally spotted a field and, as I crossed the fence at about 200 ft over what looked like a postage stamp, I shook the stick and yelled, "Full flaps." I couldn't have hit the field had I put the nose straight down. Lieutenant Cope hit the throttle and yelled, "Take me home."

On the way back, I flew very carefully, desperately trying to think of something to atone for my big goof-up. I decided I had to fly a perfect traffic pattern and make a perfect landing. Just as I started to enter the traffic pattern, he grabbed the stick and said, "I'll land it." I was sure that I had washed out. Back on the ground, he was filling out the form and hadn't said a word. I piped up, "Sir, I know that I can do better than that. Would you fly with me again tomorrow?" He looked surprised, and with the slightest hint of a grin, he said, "We'll see." I'm sure that no one had ever asked to fly with him. I never heard another word about it and continued on with my training. I'm sure my instructor put in a good word for me.

After moving on to basic flying training school at Enid, Oklahoma, we began night flying. For night flying, we would fly out to a remote auxiliary field, usually a large salt flat (dried-up lake) in northwestern Oklahoma. Portable lights were set up to mark the landing area. In 1943, many small Oklahoma towns had a double row of streetlights marking the main street and that was all. After one night flight, I returned to the field, or so I thought. As I approached touchdown, I met a car coming toward me. I was landing on the main street of a small town. Luckily, I had time to go to full throttle and get back up in time to miss the car and move on to the auxiliary field.

On my first cross-country flight, at Enid, I dropped my navigation map with my course marked on it. The map went out of reach to the bottom of the plane. After vainly trying to reach the map, it dawned on me to invert the plane and let the map fall into the canopy where I could grab it. When I inverted the plane, all the dirt and trash on the floor fell into my face, along with the map. I was finally able to grab the map before it flew into another part of the airplane.

After graduation from advanced flying school at Foster Field in Victoria, Texas, in May of 1943, I was ready to win the war all by myself. After all, I was probably the hottest new pilot in the Air Corps, self-evaluated of course. After 5 days of waiting for some choice assignment, I and four others, the last of class 43-E, were assigned to Matagorda Island Aerial Gunnery Range, where we had just spent two miserable weeks during our aerial gunnery training. Our assignment was to tow target pilots. Talk about lowdown and bitter, this was considered the armpit of the Air Corps. After a few months of swallowing our pride, it turned out to be a pretty good assignment and I learned a lot. I became a gunnery instructor and, eventually, an advanced flight instructor back at Foster Field.

After being shown the gunnery range over the Gulf of Mexico, I was assigned my first mission as a tow target pilot. The enlisted man who was the reel operator in the rear seat was a "don't give a damn" Pfc. who was constantly being busted from and promoted back to Pfc. He was very personable and knew his job well. We took off and, with a flight of cadets following, went to the gunnery range. The reel operator let the target panel and tow cable out of the bottom of the airplane.

With the weight of the hydraulic reel motor and the drag of 150 yd of steel cable trailing behind, the plane lost all stability and started lunging like a bucking bronco. I could not get it under control. The Pfc. volunteered, "Uh, sir, some of the other pilots let me handle the controls some while we are in the air. Would you like me to do it for now?" Needless to say, I answered, "Yes." Just as soon as he took the controls, the plane started flying as smooth as silk. I got the feel of the plane, took over, and completed the mission. I thanked him profusely, and it was never mentioned again. From then on, I requested him to be the operator as often as possible. I learned a lot about flying from that Pfc.

While stationed at Matagorda Island, there was absolutely nothing to do on your days off except to fish or swim a little or go to the mainland occasionally. We could check out planes and go to an auxiliary field to vent our frustrations through aerobatics, takeoffs, and landings, with a bit of mischief thrown in. Once, we flew to an unmanned auxiliary field on the mainland to play around a bit. When we were ready to return to the base on the island, we prepared to take off, but one of the planes would not maintain the proper revolutions per minute on the mag check. We had learned all the way through training never to take off if the revolutions per minute fell off during the test. The pilot was married and the father of a small child in Victoria and would not take off. I volunteered boastfully, "I will fly it back to base." As I gave it full throttle, it seemed to have the power, but as I neared the end of the field with its wire fence, there was some doubt as to whether or not it would make it. It was too late to stop and, after a bounce or two, it staggered into the air. I headed back toward the island but was never able to get it 100 feet above land and then over the waters of Matagorda Bay. When I landed safely, I vowed never again to take off without the proper revolutions per minute.

Close formation flying was a favorite activity, sometimes too close for comfort. One activity was touching wing tips in flight. Strictly a no-no. Another was flying so close that the wingman's wing was just inches behind the leader's wing and the wing tip was just inches from the leader's fuselage. In this position, the wingman's propeller tip was so close to the wing tip of the leader that the prop wash off the prop tip made it very difficult to maintain stability of the leader's plane.

Another foolish activity was for the wingman to approach the leader from above and behind, push the nose down and gain speed, start a barrel roll when in formation, and when directly above and inverted to the leader, stop the roll for a few seconds and look "up" at each other with silly grins on our faces.

One positive activity during these times was to practice rolls and inverted flight just above the smooth upper surfaces of cloud layers. The times that I failed to complete the activity successfully convinced me to never try it just above the ground or runway, tempted as I was on occasion.

A friend and I checked out a plane for a trip to New Orleans for a little rest and relaxation. On the way, we flew over the infamous swamps of western Louisiana. We dropped down below the treetop level and followed the channels of the meandering streams and bayous. Some of the turns in the streams were quite sharp and there were close calls, for the snakes and alligators, too, I imagine.

Another memorable incident occurred while I was ferrying a plane from Moore Field in south Texas to Matagorda Island, Texas. It was a very hot day in July and a long, boring flight over miles of mesquite range. I decided to remove the safety belt and squirm around a little bit. That helped some for a little while, and then I decided to relieve the monotony by doing some aerobatics. Yep. I forgot about the loose safety belt. I rolled the plane over on its back and, suddenly, was falling out of the airplane. I caught the tips of my toes behind the instrument panel and hung on. I completed the roll with the stick and then fastened the safety belt. I had my parachute on, so I was in no real physical danger, but I sure would have been when I got to Matagorda and tried to explain how I lost the airplane.

After being returned to Foster Field as an advanced single engine flight instructor, I spent a lot of time teaching my favorite activities: aerobatics and aerial combat dogfighting. While leading a flight of six in trail formation to demonstrate how to escape from a plane on your tail, I cut the throttle and cross-controlled. The plane skidded around 180°, and I gave it full throttle and headed back head-on at the flight, barely avoiding collisions with the other surprised flight members.

Another favorite activity was to play chase through the valleys of the huge cumulus clouds with other instructors on our days off. Sometimes the gaps in the clouds would close or the turns were too sharp, and we would fall into the zero visibility of the inside of the cloud, never knowing who else was staggering around inside it. A very dangerous game.

Overconfidence while firing on a towed target during P-47 transition training at Venice, Florida, in March 1943, got me into real trouble. Having been an aerial gunnery instructor, I had fired on towed targets hundreds of times and could easily score 80 to 90 hits out of 100 rounds fired. I was firing on a target over the Gulf of Mexico and could tell that most of my rounds were hitting the target. As I reached the minimum deflection angle, I kept firing. One round went in front of the target and cut the tow cable. The target stopped right in front of me, and I could not avoid it. It hit my propeller and the propeller cut the heavy metal weight that kept the panel vertical off the iron pipe to which the panel was attached, and the heavy weight went through the oil cooler. The spewing oil on the hot engine started smoking, and I headed back to the shore and Venice.

I called for an emergency landing and the field was cleared for me. As I started to turn onto the final approach, having lost all of the oil, the engine froze and the propeller stopped dead still. With the P-47's gliding ability and the frozen propeller, I had to stick the nose toward the ground to prevent a stall. I could not complete the turn, but landed on grass and crossed the runway, hit a ditch, and became airborne again. I landed again and was immediately into the field of cypress stumps left on the infield when the runways were cleared. Things got really rough when I got into the stumps. The plane broke in two behind the cockpit. All three landing gears were torn off. The wings curled and the guns in the wings were bent. The plane almost tipped over but settled back upright, and I turned off the ignition and yelled at no one, "Switch off." My only bruise was to my knees when the stick flopped around during the crash.

Upon being assigned to the 493d Fighter Squadron of the 48th Fighter Group of the Ninth Air Force, my days of fun and games ended. My aim now was to complete the mission and survive. Any daring acts, sometimes with reckless abandon, were for the purpose of doing the job to complete the mission and help win the war.

I truly believe that surviving these and many other incidents was instrumental in my development as a fighter pilot and ultimately my survival as a combat fighter pilot. I believe it is called "fighter pilot mentality."

Dropping In for a Visit

Told by Frank N. Davis, Jr.
Lieutenant Junior Grade, U.S.N.

In January of 1943, I was at Naval Air Station Olathe, Kansas, for primary flight training in the Stearman biplane. I had advanced to the aerobatic stage and was having a fabulous time doing loops, rolls, or whatever.

One day when I took off, there was a layer of clouds about 2000 ft with about eight-tenths coverage, and the sun was shining brightly above. Of course, at this stage of training, we were never to go near a cloud, but I figured I could go up through one of the holes and, with shrewd calculations regrading wind and flight direction, I would be able to practice aerobatics. After all, I could still see the ground through the holes in the clouds. So I climbed up through a hole and broke out on top in the bright sunlight eager to get started.

After about 20 minutes of exhilarating aerobatics and wringing myself and the Stearman out, all the time making expert position calculations, I decided it was time to go back down under the clouds for a look around the countryside. When I got below the clouds, nothing looked familiar, even though I thought I had never been too far from home base. Again, with more expert navigation calculations as to what had gone wrong, I headed in two different directions before realizing that I was lost. I knew I could not fly around for too long on the available fuel and was beginning to think that I might be in big trouble. My decision was to land and take it from there. Looking around, I saw a large pasture with cattle grazing, and it looked better than the smaller plowed fields in the area. I determined the wind direction as per training and came in for landing among the cattle with a sliding stop just before splintering the back door of the farmhouse.

The resident farmer rushed out to see what was going on. After all, it is not every day that an airplane drops in for a visit. I climbed out of the cockpit leaving the engine running, as I could not restart it by myself. We met and I asked the farmer where I was and he replied, "You must be from Olathe." I said, "Yes, and which way is it?" Then he informed me that it was about 40 mi straight north up the highway in front of his house. I thanked him and crawled back in my trusty Stearman. By this time, I had audiences in the house's yard and on the highway in front of the house.

In planning my takeoff, all of a sudden it occurred to me that I could not take off into the wind because of all the large trees around the house and along the highway. I had just barely gotten into the field on landing. Therefore, I would have to take off downwind. I taxied up near to the front fence and trees, swung the tail around, applied full throttle, and went charging back again through the cattle on my runway of pasture, mud, and dung. I was building up speed with the tail off the ground approaching the back fence, but the bird was rolling on the front wheels holding to the ground, solid as a Rolls Royce. Regardless of whatever, I was committed and on the way. Just before I got to the fence, I pulled back on the stick and had barely enough speed for lift to mush over the fence and plop down on the other side. Then I had a whole new field to finish my takeoff run.

After getting airborne, I went back to the highway, waved to my audience, headed due north, and soon Naval Air Station Olathe appeared on the horizon. After landing, as the plane captain chocked the wheels, he asked what I had done to his airplane with all the mud and debris underneath the lower wing. I told him I ran off the taxiway, which was true. I was about 40 mi off. I went back to the hangar and did not tell anyone about my episode and hoped that my audience down on the farm or the plane captain would not start any vicious rumors to prompt an investigation. Silence was maintained until after graduation in September of that year, and then it was one of my favorite stories at flyboys' get-togethers.

My First Mission
Was Almost
My Last

Told by Charles R. Mott
Captain, U.S.A.A.F.

While I have always believed I had a successful career during my years of military service, it almost ended before it began. Upon completion of my flight training, I was sent to an air base in San Francisco, California. Our squadron was assigned the task of patrolling the West Coast in order to be on the lookout for Japanese submarines. The plane I flew was a P-38. One day, while on a routine patrol of the coast, I got an idea. I decided that, for a little additional excitement, I would fly under the Golden Gate Bridge, certainly one of the most famous bridges in America. I successfully navigated my plane under the bridge and returned to base feeling like a real ace. What I didn't know was my antics had been reported. I was met by my commanding officer who let loose a few expletives about how reckless and irresponsible my stunt had been. Needless to say, my feelings regarding my superior flying techniques vanished under the verbal assault I received. Had there not been a desperate need for fighter pilots, I might have been court-martialed. Instead, the punishment was to be my immediate departure overseas.

The trip overseas was by ship and I arrived in North Africa in February of 1943. It was well after dark when we finally disembarked. What followed was a long and tiring hike to our quarters. It seemed we walked half the night. After several hours, we arrived. Although designated as our campsite, nothing was there: no tents, cots, or supplies. I momentarily felt better when informed that the officers would be staying in a separate location. Again we walked and walked. When we arrived, again there was nothing: no tents, cots, or supplies. The only thing different about the officer quarters was a sign indicating the area was for officers only!

As I stood in the dark, I could see a small light off in the distance. Discovering what this light was just might be worth the trip. Again, I set out walking. Despite walking in the direction of the light, I didn't seem to be making much progress. The light wasn't getting any brighter. Finally, after walking for what seemed like hours, I arrived at the light. What I saw made me fall down laughing. Standing in the tent was a young soldier holding a candle in one hand and a stick with a strip of bacon in the other! This guy was very optimistic or very patient. Either way, I sure hoped he wasn't going to be our cook.

Thus began my overseas career. I spent 19 months in Africa, Sicily, Italy, and China before returning home to Tennessee in August of 1944. I was promoted to captain in December of 1945 and discharged the next month. I never considered my military career anything but a great honor. I loved flying and was proud to have served my country during the crisis which we call World War II.

How Not to
Become a
Great Hero

Told by Enoch B. Stevenson, Jr.
Major, U.S.A.A.F.

The day was to become a great opportunity. It started with the usual briefing for escort over the Continent in our P-51 Mustangs. Everything was routine until we reached the target area where the weather was lousy, as was often the case. As it happened rather frequently, separation took its toll, and I found myself alone, me and about forty 109s and 190s. They were well above me and silhouetted against a gray overcast.

Visions of heroism and its rewards—Distinguished Service Cross, Medal of Honor, and ace in one mission. You name it, and I had already thought of it. I added power and began my climb to take on Tail End Charlie. I could just keep working my way up the formation until my ammunition was exhausted or I was. Ah, the best laid plans.

When we began our combat tour, fighter command sent Lieutenant Colonel Tommy Hayes to brief, lead, and impart some of his wisdom gained in combat. Among his other talents, Tommy could draw well. One afternoon in the officers' club, he showed a couple of us that the fuel pump of the 109 was located in the root of the left wing, and if you got him there, you had him. If he didn't flame, at least his engine would quit. What a wonderful opportunity to put that bit of wisdom into practical application.

Unfortunately, I got a first-class case of buck fever. When I got close enough to recognize the 109, I started firing, but as any fighter pilot knows, I was probably miles too far out of range. Suffice it to say, I wasted my ammunition. I did manage to get one confirmed, but I didn't get the other rewards and awards I had in mind.

My First Mission

Told by William A. Potts
Lieutenant, U.S.A.A.F.

My squadron was formed in California. Our route to over-seas service took us from Bakersfield, California, to Cairo, Egypt, and Italy. My military service started with Company E, 117th Infantry, 30th Division, Tennessee National Guard stationed in Dickson, Tennessee. Of this I am very proud. We were federalized in 1940.

I later became an aviation cadet, class of 43-F, volunteered for night fighters, and went with my squadron to North Africa on our way to Poltva, Russia, to give cover to B-17s landing and taking off during the triangle bombing missions. We were stopped in Cairo, Egypt, at the pleasure of Stalin. Then we proceeded to Italy. We were told the Russians did not want more Americans there.

Our airplane was the P-61 Northrop Black Widow, the largest designed fighter in the Army Air Corps during World War II. The greatest things pilots have going for them are equipment and training. We currently had both. Later, I was assigned to other aircraft.

My first mission didn't mean much for the war effort, but it was a great experience for me—the best I can remember after more than 50 years. I could entitle it, "This, That, and the Other, and Things Like That." It went like this.

Soon after my arrival in southern Italy, the war was slowing down as far as the German Luftwaffe was concerned. I think their air effort was in France and further east of Germany. I was scrambled to intercept a German JU-88. They were used for reconnaissance over Naples, Italy harbor, and the area. The Germans were in need of knowledge of our troop activity and shipping, as the Italian campaign was getting very serious for them. After getting airborne, the ground control station sent me scouring northward in pursuit of the picture-taking German airplane. I would have been delighted to have prevented him from getting the pictures back home. But after getting completely lost, I was advised to give up chase. My intended target had a big head start and was gone.

Now my task was to go home, but where was I? So I did the normal thing and called the ground control station for a heading, but there was no answer. After about the third call, I heard this perfect English-speaking voice advise me to take a heading of 290° or so. He was trying to lure me further north from my home base and finally getting me to land at their field. It was a great midwestern voice, like maybe he was from Lincoln, Nebraska, Sioux Falls, South Dakota, or someplace like that, definitely not the Bronx or Plantation, Mississippi.

Well, I reasoned that a person whose relatives came from Raney Camp Hollow of Northwest Dickson County, Tennessee, should know that 290° was going north. Italy runs northwest-southeast, so I needed a southern heading. After he gave me this instruction three or four times, I gave him a few words and that was the end of his transmission. I guess the German pilot was by now in Austria, Greece, or someplace laughing at me.

I flew south awhile, and finally my ground control station came in and gave me directions home after asking, "Where have you been?" My answer, "I don't know." I landed back at the base safely with all my ammo and a lot less gasoline.

Now, if you were a taxpayer in the United States at this time in World War II, I'm sorry my first mission didn't have better results for you. But I can now report that I did better later. I think war is killing people and breaking things. I hope and pray that it never happens like this again.

Midair

Told by E. H. Bayers
Commander, U.S.N.

In the spring of 1943, I was assigned to NAS Melbourne, Florida, as an advanced fighter pilot instructor in F6F Hellcats. At that point in time, each instructor was assigned five students, two 3-plane divisions. In the course of instruction, the students were taught night carrier landing breakups. Approaching the active runway, the first section went into a right echelon while the second section remained intact. When the leader of the first section separated for his approach, the second section leader was to follow him around the runway and then go into a right echelon for their breakup.

On this particular night, I was leading the two divisions to the field for a normal carrier breakup. As I entered final approach, wheels and flaps down, apparently the section leader of the second division lost sight of me. He collided with my plane from the port quarter, slightly below me. His cockpit enclosure struck the bottom of my plane, causing him to veer up. My propeller cut his plane in two, causing him to crash. When my propeller struck his plane, it was damaged to a point where I could not use power due to excessive vibration.

Experiencing the vibration, it was apparent that I was in a plane without power. I tripped my safety harness and stood up to bail out, at which time I discovered that I was too low to parachute safely. I sat down again in the airplane. The runway that I approached had a drainage ditch in front of it which I had to clear to make a landing. I maintained the glide, barely clearing the ditch. I hauled back on the stick and squashed the airplane to the ground. The airplane did not roll but sort of bounced, which was extremely fortunate for me. When I viewed the airplane the next morning, I discovered that both wheels had been turned 90° to the line of flight. If I had the opportunity to make a normal landing, the damage to the wheels would have caused the airplane to nose over. This event would probably have caused my death, as my safety harness had not been reconnected. Both my airplane and those of the second division wingmen had been severely damaged by the accident.

Becoming a
Fighter Pilot

Told by Robert F. Hahn
Major, U.S.A.A.F.

My fire to be a fighter pilot was ignited when I saw color pictures of the Army P-35. I did apprentice work (free) because I enjoyed being near them. I worked on a farm on weekends to earn enough to pay for my mother and maternal grandmother (then only 80) to get a flight in a Ford Trimotor many years before my first flight in primary.

Sent to the classification center in Nashville in 1943 (100 Oaks area), I was classified as a fighter pilot. We would be marched to a local swimming pool, where I met a pretty redhead. Though we never had a date, we corresponded. I was so impressed with her that I put her name on my P-51 Mustang in England (50 years later, we are now married).

After Nashville, via Maxwell, my primary flight training was at the Mississippi Institute of Aeronautics (MIA) near Jackson. Perhaps most pilots flew only one type of aircraft in each phase of flight training, but not in my case. My instructor, James Paisley from Ohio, first soloed me in the Stearman (had no problem, never ground looped). Then, after the specified hours, I was also required to solo in the PT-19, which was easy compared to the Stearman. Basic was in the Vultee Vibrator BT-13 at Greenville, Mississippi. Then on to advanced in the AT-6 at the Jackson, Mississippi, municipal airport.

My graduation wings were quickly taken to my Nashville sweetheart (she showed them to me recently). Then on to Eglin for ground and aerial gunnery in the T-6, prior to an extensive fun checkout in the P-40 at several north Florida fields and before a fast trip to Liverpool, England, on the *Mauritania*, sister ship of the ill-fated *Lusitania*.

With less than 4 hours flight time in north England in a P-51B Mustang using a shotgun shell to start the engine, I finally arrived at Raydon, East Anglia, England (home of the 351st Fighter Squadron, 353d Fighter Group). On to my first frightening mass formation flight from England over Germany, I escorted B-17s.

Many missions were strictly escort (B-17 or B-24), picking up our bomber group at landfall (across the English Channel), and staying with them until they returned to the Channel, whether or not we engaged enemy fighters. Our flight altitude was generally 30,000 to 36,000 ft, but sometimes lower. My longest mission from takeoff to touchdown (without action) was 8 hours. The P-51 Mustang was truly a long-range fighter.

There were also many missions on which we escorted our own bombers past the target and another group of fighters would escort them home. Then we would drop down for targets of opportunity or, at times, assigned targets on the ground: airfields with open or dispersed aircraft, flak towers (sure to raise your blood pressure), military vehicles, and railroad marshaling yards and locomotives.

My first victory was an ME-109. Call it "first time" emotions, poor coordination, or poor gun-bore sighting, my bullets from the six guns were just spraying the other aircraft, which apparently frightened the German pilot into bailing out. Being very chagrined that I was not seeing both plane and pilot going down in flames, I chopped the throttle and did some quick maneuvers to get the free-falling pilot in sight. When his chute blossomed, I got him centered on my nose and was about to fire when I thought, "No! I'll get him next time." My No. 2 man was still nearby. We climbed up and found Nos. 3 and 4 and resumed escort.

It was very pleasing to be selected to fly the group commander's wing, which was my position for six missions with Berlin as the target. My Mustang was fitted with a special camera aimed down between my right wing and tail. At an appropriate time before the bombers' "bombs away," the colonel signaled me to leave and I descended from about 30,000 ft to 1500 ft. During the rapid descent, I had to deal with engine temperature and other things. It would have been nice to have speed brakes as we did on the faster jets. Of course, I did not think of the hazard of the 108 gal of 150-octane fuel in the external fuel tank. With the aircraft slowed to the bomber's speed and at 1500 ft or less, I watched the falling bombs and positioned my aircraft to get pictures of the bombs hitting their targets from start to finish. Then I turned my camera off and climbed. Amazingly, with only one set of eyes and head on a swivel, no radar, and no homing devices, it wasn't long before I was at altitude and back on the group commander's wing.

My fifth aircraft downed was also an ME-109. It was an easy one with altitude and position advantage. My gunfire cut off the right wing, and he spiraled downward. Climbing back above the bombers with No. 2, we saw an unusual sight above us: 39 ME-262s were in show formation, 13 three-ship Vs, line abreast. They were soon out of sight. Back at the base in mission debriefing, there were several bomber and fighter groups that reported sighting this unusual formation. I said to my No. 2 pilot, "It would have been so unfair of us to attack them, two experienced U.S. pilots against only 39 German jet students."

In and Out
of the Clouds

Told by Bill Burch
Lieutenant, U.S.N.

It was May 21, 1943. We were in the South Pacific in the vicinity of New Georgia. I was flying lead, second section, in a four-plane division of F4F Wildcats. We were at 25,000 ft when we ran into extremely heavy clouds. Following standard procedure, I turned 30° to starboard after losing sight of the flight leader. Apparently, my wingman did the same thing since I was unable to see him. After flying on instruments for about 10 minutes and being unable to see anything but thick clouds, I decided to drop down to see if I could get in the clear.

Finally, at about 21,000 ft, I came into the clear. To my surprise, about 30 seconds after breaking into the clear, I saw a Japanese Zero just below and ahead of me. I pulled up into the clouds again hoping he hadn't seen me. I didn't have my gun switches on or my gun sight set. I remedied that at once and was ready for the Zero. I pushed the throttle forward three or four notches and eased down out of the clouds again. This time, I was ready and started a run as soon as I saw him. The fuel tanks of a Zero were in the wings next to the cockpit, and that is where I aimed as I made my run. I don't think he ever saw me as I made my run from the stern. I was aiming a little far ahead with my first burst, but the second went right into the root of his left wing, and an explosion ripped the plane immediately. It was my very first smoke.

I reached home base at Guadalcanal about 20 minutes later without my wingman. After landing and reaching the operations tent, I found the other three members of my flight already back. I received credit for a probable smoke because there were no witnesses and no gun camera.

Bogie

Told by Lee V. Gossick
Captain, U.S.A.A.F.

This story is not about any particular mission I flew during my World War II time in North Africa and Sicily. It is instead about one of the many and diverse things that come about under combat conditions and add some humor to day-to-day life, which for the most part is far from humorous. During the summer of 1943, our unit, the 79th Fighter Group, was operating out of an airstrip in the central part of Sicily. It was a dirt strip, carved out of a Sicilian farmer's wheat field by Army bulldozers, with taxiways leading to the dispersed individual parking spots for our P-40-F Warhawks. While working and living conditions were quite primitive, they were much improved over the conditions we had experienced in the North African desert. Groves of orange and almond trees, springs with clean water, and a generally friendly populace were a much welcomed change from the barren, dry, hot, and sandstorm-ridden environment in which we had operated since leaving Egypt in late 1942.

In August 1943, our group was flying dive-bombing and strafing missions in support of the British Eighth Army's advance up the east coast of Sicily. General Patton's forces had moved toward the northwest and northern sectors, and a race was on between General Montgomery and General Patton to reach the Messina Straits and cut off any large evacuation of German forces to Italy. From our airstrip, we were within sight of Mt. Etna, with its wispy, vaporous reminders of its potential for creating far more destruction than anything our weaponry could do. We flew missions as fast as the ground crews could refuel and rearm our aircraft. Three and four sorties each day per pilot were not unusual. Our targets were primarily anything moving north and east, plus the boats, barges, and any other craft involved in the German attempt to withdraw across the Messina Straits. As the German forces were pressed into the area, the concentration of antiaircraft weapons became so intense that strikes against these targets were frequently paid for by losses of our pilots and aircraft.

Shortly after General Patton's forces had taken and secured the city of Palermo on the northwest coast, our squadron's engineering officer, Lieutenant Tony Cirrito, requested permission to take a Jeep to go see if he could find his uncle who lived in Palermo. Permission was granted since we were in somewhat of a lull in operations because of weather conditions. After a couple of days he returned. He had found his uncle safe and well. He had insisted on sending Tony back with a very large, gray, domestic goose. According to Tony's explanation, his uncle wished the goose to be the mascot of our squadron. But I believe Tony had other thoughts in mind. Somehow, early on in my time with the squadron, I had picked up the nickname of "Goose." I believe there was at that time a rather famous baseball player by the name of Goose Goslin. Or maybe it was the way I walked. Nevertheless, the nickname stuck and I ended up using it in naming my P-40 *The Gruesome Goose.*

As you might guess, I ended up as the custodian and keeper of the goose. We didn't know whether we had a goose or a gander, so the name "Bogie" (fighter pilot terminology for unidentified aircraft) seemed appropriate. Bogie was a very friendly and trusting bird and always on the make for a handout. I was somewhat concerned that this big, fat goose might be too much of a temptation for some of the men who had been on K rations for the past year. But shortly after Bogie's arrival, there was an incident that made it seem unlikely that anyone in the squadron would do Bogie any harm.

One morning as the squadron commander, the operations officer, the enlisted operations clerk, and I were at work in the operations tent, we were visited by a colonel from higher head-quarters who was giving us a royal chewing out about the rather casual dress and lack of military courtesy he had observed as he made a walk-through inspection of the squadron. The colonel was impeccable in his clean and freshly pressed uniform, and his shoes were shined to the West Point standard. As Major Uhrich, the squadron commander, the operations clerk, and I were standing at stiff attention, Bogie quietly waddled into the tent unnoticed by the colonel, whose back was to the tent opening. The lecture came to an abrupt end as Bogie quietly positioned his rear end squarely over the colonel's right shoe and let go in the mode that lies behind the phrase "loose as a goose." As the colonel exploded, Bogie made a hasty exit, narrowly escaping an angry kick by the newly dec-orated shoe. The story spread quickly among the troops, proba-bly embellished somewhat with each telling. From that day on, Bogie could do no wrong, and I had little concern about his becoming a "roast goose" item on the mess tent's menu.

Bogie seemed interested in the sight and sounds of our P-40s warming up, taking off on the missions, and their land-ings upon return. His favorite scrounging area was around the operations tent where missions were scheduled and where the assigned pilots were briefed and debriefed after returning from their missions. The pilots and ground crew members were easy marks for Bogie's polite but persistent panhandling, and his appreciation for such goodies as a bit of chocolate or a cracker was usually demonstrated by a vigorous side-to-side shuffle of his tail feathers.

As I would leave on a mission, I would pick Bogie up and place him on the hood of the Jeep that carried three or four pilots to the dispersed parking spots where our aircraft had been readied by our ground crews. Arriving at my P-40, I would put Bogie up on the left wing tip where he would sit while I was putting on my parachute and getting strapped into the cockpit. Upon hitting the starting switch and at the first sound of the big Packard engine coming to life, he would leap off the wing tip and settle in for a snooze in the shade of an oil barrel or other equipment kept in the parking area. Bogie would invariably stay there until I returned from a mission, which on one occasion entailed some 7 hours, as I had experienced a hydraulic problem causing me to divert to another airstrip for repairs before returning.

To my knowledge, European domestic geese are not much when it comes to flying, particularly when they are overfed, as was the case with Bogie. Nevertheless, inspired perhaps by all the flight activity around him, he would occasionally make a stab at it. In the evenings, after the last missions had been flown, the squadron members would gather around the operations tent to rehash the day's action and contemplate what was likely to be in store for the next day. Sometimes, when the spirit moved him, we would see and hear Bogie charging down the Jeep path, wings flapping furiously and honking loudly, presumably to ensure our attention. But alas, his lift-to-weight ratio was not in his favor, and he would reluctantly give it up for the day.

But on one fine evening, the gods were in Bogie's corner. A stiff breeze was blowing directly up the path, and as he once more tried it, he found himself airborne! He flew by the operations tent at an altitude of some 10 or 12 ft, honking triumphantly. I believe many of us watching his effort would have sworn that between honks Bogie was grinning with delight! He continued on for perhaps another hundred yards down over the wheat field when he apparently realized it was going to be a long walk back. He was attempting to turn back in our direction when he literally stalled out, disappearing in a cloud of dust, wheat chaff, and feathers. It was a good 30 minutes before he trudged his way back to the applause of our group, dusty and feathers somewhat disheveled, but obviously very pleased with himself.

Not long after this episode, having completed my combat tour of duty, I was ordered to return to the States. The 79th Fighter Group was in the process of moving from Sicily to Italy, but so far as I know, Bogie was not listed on the movement orders. I can only hope he found the companionship of some other Sicilian geese and lived a long and pleasant life.

Messerschmitt 262

Told by Clifford J. Harrison, Jr.
First Lieutenant, U.S.A.A.F.

The ME-262 was the first and only successful jet fighter plane to get into combat in World War II. It was fortunate for the Allies that it did not get into production earlier and that, when it did get into action, there was not enough fuel to make more adequate use of it.

I saw ME-262s on two occasions. The first time, our flight was heading back from the Crailsheim area. My flight leader had been hit and his plane was on fire where the belly tank had been. He had jettisoned his bubble canopy and was about to bail out when we told him the fire had gone out. I was flying his wing and another element of P-47s was with us. An aircraft in the distance that we finally identified as an ME-262 turned toward us. Our second element turned into the 262. The German plane came within about 200 yd of us, then pulled away, and was gone in seconds. It was the fastest airplane I had ever seen.

My second ME-262 sighting came when we were again returning from a close support mission. We saw what looked like an aircraft on the autobahn. We went down for a closer look and found it was an ME-262 that had apparently made a crash landing. We also discovered that in a wooded area beside the highway there were many more 262s parked among the trees.

We immediately began strafing runs on the 262s. Several other flights in the area heard us and joined the action. It was not long before P-47s were strafing 262s in every direction— the fighter bomber equivalent to the famous Navy turkey shoot. It was amazing that there were no collisions, but I know of at least one near collision. I was on a strafing run with a 262 in my sights when I noticed tracer bullets coming from above and behind me. Suddenly, a P-47 flew within feet of my airplane. It seemed to me that the airplane flew between my engine and my right wing, shooting all the way. I received a tremendous jolt from its prop wash. I saw that it was the airplane flown by one of my roommates. When I returned to base, I more than casually mentioned the close call. I found that, if I hadn't told him, he would never have known about it since he never saw me.

One footnote to the near collision. On a previous mission, the same pilot had made a strafing run 90° to my run. I was following our flight leader, so I know my pattern was correct. Our P-47s crossed seconds apart, similar to what you sometimes see at military airshows.

A couple of years ago, when I was visiting my son in Minneapolis, I telephoned this pilot, my former roommate. I had an 86th Fighter Group Association roster and knew his address and phone number. I had not seen or talked to him since the war. When his voice came on the line, without any other comment, I said, "Parkgate Gold 4, quit screwing up the strafing pattern before you kill somebody."

There was a long period of chuckling.

Southwest Pacific— June 30, 1943

*Told by Bill Burch
Lieutenant, U.S.N.*

Guadalcanal, Solomon Islands—southwest Pacific—8850 mi from the United States—10,893 mi from home in Tennessee. Temperature 101 degrees, humidity 89 percent, insects swarming, foxholes bare and damp. Fighter One Airstrip—Navy Fighter Squadron 21, "Blackjack Squadron." CO Whitney Ostroiyi, XO Herc Henry.

Enemy—Japanese—daily scraps. Had three kills to my credit before June 30. Aircraft—Navy F4F—Wildcat. Top speed 420 kn, 1850 hp. Six .50-caliber machine guns, three in each wing, 1500 rounds of ammunition for each gun.

Combat air patrol begins at daybreak and ends at dark. Length of patrol 3 hours before relief. Had the early patrol on this day, takeoff at 0500. Station 10,000 ft over Russell Island about 30 mi off Guadalcanal. 0601—Cactus Control to Black Jack Leader—large formation of bogies 40 mi west— 12,000 ft. Your course 278° for intercept—"Buster."

We climbed to 15,000 ft on the heading given and soon spotted six Japanese medium bombers escorted by 12 Zero fighters. They were headed for Cactus. We called Cactus and told them our four-plane division would need some help.

We prepared to attack—gun switches on, revolutions per minute high, fuel rich, sun goggles down, and hand on trigger. We got in position for a high-side run on the three nearest bombers. I rocked my wings and peeled off to make a run, and each of the other three Wildcats followed one by one. I could see tracers going into the outside bomber as I made my run, and it was smoking good by the time all had made their runs. We started to climb back into position to make another run when the Japanese fighters started a run on us. We were in an awkward position since we were climbing and our speed had dropped considerably. Fortunately, we spotted some clouds to the north and slightly below us. We dived for the clouds and the first three of us made it okay. However, Tail End Charlie, our No. 4, was hit and didn't make it. We sneaked out of the cloud formation and made a low-side run on the bombers again (now five of them). We were able to smoke one more bomber before retiring to the clouds again. By that time, our fuel was getting low and we got down close to the water and island-hopped back to Fighter One.

An Unusual
Experience

Told by J. Pat Maxwell
Major, U.S.A.A.F.

After graduation from flying school in December 1942 and being commissioned a second lieutenant in the Army Air Corps, I completed P-47 fighter training and was transferred to England. There I was privileged to be assigned to the 78th Fighter Group, 84th Fighter Squadron from April 1943 until January 1945. During that period, I flew 133 combat missions with 432 combat hours in the European theater of operations.

The 78th Fighter Group, equipped with P-47s, was the second fighter group in the Eighth Air Force to go operational in World War II in the European theater. This unit was preceded only by the Fourth Fighter Group which was formed with American pilots already in England as members of the RAF Eagle Squadron and who had participated in the Battle of Britain.

The 78th Fighter Group was stationed at Duxford, England. The group was operational from April 1943 to April 1945. It flew 450 combat missions, destroyed a total of 697 enemy aircraft while suffering the loss of 167 aircraft, and produced the first ace in the European theater of operations. Its primary mission was to provide escort for the Allied bombers during daylight operations and to conduct dive-bombing and strafing missions as well as ground support for advancing Allied troops.

Reflecting on my own personal experiences, including air-to-air combat, one of my most bizarre experiences occurred on my 15th mission. The date was July 1, 1943. The mission, called a fighter sweep, called for 36 P-47s to fly from England over the North Sea, crossing Belgium and other enemy territory in an effort to engage enemy aircraft. I was flying as a wingman in a flight of four P-47s. My job was to maintain position in formation with my leader to protect him from enemy attack and to assist in spotting enemy aircraft.

Shortly after the group crossed over the Belgian coast at 29,000 ft, unidentified aircraft were spotted to the south. The group turned south to investigate. The unidentified aircraft turned out to be a large group of FW-190s. The fight was on! My flight leader rolled over and closed on an aircraft right below us. The FW did a half roll into a steep dive. My flight leader followed right on his tail. In order to stay in position on my flight leader, I had to roll into a split-S. When my aircraft reached a vertical position in the dive, I experienced a condition called "compressibility." This was a condition experienced by the heavy P-47 in a steep dive above 20,000 ft. While experiencing this condition, the flight controls became ineffective. The only way to overcome this condition was to slow the speed by reducing the throttle and ride it down until the aircraft reached a lower altitude where the controls became effective. My particular problem was compounded because of our high initial altitude and the high airspeed I had reached in the dive.

When I attempted to reduce the throttle, the nose of the aircraft tended to go past the vertical, and buffeting occurred. I tried to use the trim in an effort to remain vertical and ride it out. With feet braced, pulling on the control stick with all of my strength, at some point after I had passed through 8000 ft, I accidentally pulled the gun switch and all eight .50-caliber guns began to fire. As this occurred, the control stick began to give just a bit. Before I could get the trim adjusted, the aircraft came up very sharply and I could not avoid blacking out. I do not know the elapsed time, but before I could see again, the canopy was completely frosted and I was hanging by my shoulder straps. I finally righted the aircraft on instruments and got the windshield defrosted. At that point, I was at 4000 ft at an airspeed of 300 mi/h. I could see that the air battle was still in progress above. My radio was out and I could not contact my flight leader. Being alone, with flak bursting around me, I dove for the deck and set a course for home over the North Sea. I reached the English coast and landed at the first strip because I was low on fuel.

The aircraft had wrinkles in the skin and a number of skin burns. The Republic Aircraft representative concluded that I had reached a speed in excess of 800 mi/h. This created some publicity that I did not particularly cherish. I was just thankful to have survived. My flight leader had also experienced some compressibility and only got credit for damaging the FW. The group, however, destroyed 10 enemy aircraft in that battle, losing only one aircraft, the group commander.

A Fighter Pilot's
First Victory

Told by Charles R. Mott
Captain, U.S.A.A.F.

The date was early September 1943. The 58th Fighter Squadron of the 33d Fighter Group moved to Italy during the Salerno invasion sometime after the initial assault by the American Fifth Army. At the time, we were equipped with Curtiss P-40 Warhawks.

Sometime prior to our arrival at Salerno, the Army Corps of Engineers had a lieutenant colonel locate a site for an airstrip. A sergeant from the corps scraped out a strip with a bulldozer under the supervision of the lieutenant colonel. This was begun in the early morning hours of the day of the invasion, and we landed on this hastily prepared strip at 1600 on the same day.

After our landing, the limited number of our crews rearmed our wing guns and added 500-lb bombs. We quickly prepared to bomb and strafe the German positions only a few miles from the end of our strip. We flew four-plane missions, four planes in each flight under the command of a flight leader. We then returned and again quickly rearmed and took off with the same mission objective of strafing and bombing German infantry, artillery, and armor positions. We continued this until darkness set in and we could no longer see to fly and pick our targets.

I don't remember much about that evening, such as where I slept, what I ate for supper, or any details of this sort. However, sometime around 2100 or 2200, our squadron commander called all the pilots and crew chiefs together. He informed us the Salerno invasion was in extreme danger because German armor had broken through the American front lines in just the past few hours. It was a possibility we might have to evacuate to the beaches and make a stand. Of course, this news was met with much conversation, many questions, and consideration of all the options the pilots would have with their planes.

As I remember, we all decided we would take our chances on the beach, and I remember hoping to be able to get a carbine or M-1 rifle since I only had a .45 automatic M-1911 pistol, which was standard issue for fighter pilots. I drank another quart of coffee, smoked at least 10 cigarettes, and tried to sleep in my bedroll.

Sometime around 0200 or 0300, an alarm sounded. It was a gas alarm. None of us pilots had kept our gas masks, and as soon as I was aware that the alarm was for gas, I went scrambling back to my bedroll. I grabbed a big cotton towel and ran, sprinted actually, to a creek that I knew was nearby. In the darkness, I jumped from the bank to the middle of the creek and landed in mud up to my knees. All I had on were my undershorts, a pair of flying boots, a towel, and my .45 pistol strapped to my waist. I couldn't move! As I surveyed my situation, I decided if I was going to die from gas, it might as well be here. If gas was present, I was merely accepting the inevitable.

I don't know how long I stayed in this position, but we hollered back and forth to each other up and down the creek. Some of the guys found a little more water than I had. Finally, we realized we were not gassed, came together in huddled groups, and got back to the squadron tent to find out what had happened. The alarm had been triggered when one of our ships in the harbor was hit by a German bomber that night. When its cargo of munitions exploded, many gas alarms on the fenders of Army trucks went off on the beachhead. This was my first night in Italy at the Salerno invasion. Fortunately, the entire 82d Airborne Division was air dropped into the area the next day, thus halting the German breakthrough at the Salerno beachhead.

The next morning, we all assembled at the squadron operations tent to be given our mission for the day. I was assigned to a flight to patrol over the beachhead area with specific instructions to prevent any attacking German aircraft from reaching the Navy ships anchored just off the Salerno beachhead. Since I had joined the 58th Squadron in May of 1943, I was a relatively new pilot with only a few missions of experience. I flew Lieutenant Bishop's, our flight leader, wing. We took off and proceeded to climb to 12,000 ft. Our four planes flew an elliptical flight pattern over the entire invading force of naval ships that had brought the American Fifth Army to Salerno.

We had flown our patrol pattern probably 10 to 15 minutes when we were alerted by radio that a flight of Focke-Wulf-190s was headed toward our area. This was the best fighter aircraft of the German Air Force during World War II. It was superior in maneuverability, speed, and other factors that made it a tough enemy when compared to the P-40s we were flying.

As soon as we realized we were in for a dogfight, my flight leader increased his engine speed to full power and started a shallow dive to gain airspeed along with the increased power settings. Trying to follow my training and experience, my head started making rapid 360° circles while I was straining to maintain my position in the flight formation.

As you can see, I was busy. We were headed south and all of a sudden eight FW-190s started dive-bombing runs from about 12,000 ft in a west-to-east pattern, so when the bomb run was over they would be quickly over their lines into friendly territory. Our flight leader came on the radio and said, "We might get one or two at the end of their bomb run, but their speed will probably prevent us from doing much more. Let's continue to dive, pick up speed, and see if we can see where they are coming out of their bomb run and cut them off."

Lieutenant Bishop's judgment was perfect. He and I got on the tail of one of the FW-190s just as they came out of the bomb run. Their speed must have been well over 450 mi/h, and we were doing about 400 mi/h. Bishop and I rolled in behind one and started firing. We saw smoke coming from the German plane, and we both gave him good solid blasts. I continued to fire even after Lieutenant Bishop had quit because I was well to Bishop's right and could keep my gun sights on the FW-190 with a few more bursts. The German peeled off to the right and went into the ground. Bishop and I pulled up sharply with heavy G forces. We were still traveling well over 400 mi/h.

The flight regrouped. Bishop, the other two planes in our flight, and I proceeded to land at our airstrip as we had exhausted almost all our ammunition. We taxied back to our parking areas, and of course, the ground crews had overheard the radio conversations and saw much of our maneuvers to get on the tail of the German fighter. We went to the briefing tent to be debriefed by the squadron intelligence officer and report our mission.

I was in many other dogfights, especially over the Anzio invasion in Italy, but I will long remember this first victory that I shared with Lieutenant Bishop, who received his orders to proceed back to the United States the next day. His tour of combat duty was finished, and mine was just beginning.

The Great Beer Caper

Told by Edward F. Jones
Captain, U.S.A.A.F.

Combat is what it is: people trying to kill each other. But all combat service was not white knuckled and clinch jawed, especially with the young pilots. We were first of all young, for the most part single, proud to be off the bench and in the game, and with sufficient downtime between missions to look for trouble to stir up.

For some unknown reason, beer was in short supply in our rations; there were only three cans for two people every 10 days. The bar stayed well stocked with whiskey; there was the combat whiskey issued by Uncle Sam to pilots on a per mission basis and scotch that we flew in from Egypt when we could borrow a cargo plane. So when word got out that the Army engineers had restored and had operational a brewery in Sicily, our inventive minds went to work. A cooperative crew chief cut openings and installed hinged doors on a set of wing tanks and a belly tank. These were filled with straw for packing purposes and off went a pilot to Sicily for the first of several beer runs. Ice was also in short supply, so on the return trip the beer pilot would go up to 30,000 ft and circle for an hour or so to get the brew good and cold. He would radio when he started down so we could get close to the runway and quaff the suds before they lost their chill.

Encounters with German ME-109s

Told by J. Pat Maxwell
Major, U.S.A.A.F.

My introduction to air combat came rather early in my tour as I was involved in dogfights with German fighters on my fourth, sixth, and eighth missions. The first was on October 24, 1943, a rhubarb that was on the deck strafing targets of opportunity in the Rome area. We were based at that time on the Salerno beachhead, which is south of Naples. Our first target was a railroad engine and four cars which were strafed; then a JU-88 was strafed and destroyed on the ground. Seven ME-109s were encountered in the Lake Bolsano area. Three were destroyed, with one damaged and claimed as a probable. They were carrying belly tanks and tried to jettison them as they were going down. They were at a very low altitude when we jumped them. We observed other planes on the airdrome by the lake. There was accurate light and heavy flak over the airdrome.

The fight on October 28, 1943, took place on another rhubarb up the west coast of Italy north of Civita Castellana, again covering the road network in the Rome area. Two seaplanes were first strafed and destroyed. Four 109s were spotted and attacked. After a dogfight, all four had been shot down. Three crashed on the same field.

After re-forming, we headed south toward the coast and observed 15 ME-109s in the air near an airdrome in the area. There were 30 to 50 multiengine and single-engine fighters on the field, some of which were taking off. A dogfight started and lasted about 15 minutes. We were on the deck at this time not over 100 ft up and in a vertical bank. Turning back across the field, I spotted a machine gun in a sandbagged hole on the field with three men jumping in the hole to get at the gun. This was the only time in 108 missions that I actually saw people I was shooting at go down from being hit by my fire. After the fight was over, one ME-109 had been destroyed, one probably destroyed, and three damaged. The others were flying singly without any kind of formation, being scattered all over the sky. Once the fight started, they would sometimes attack in pairs. The A-36 proved to be superior in turning. Time and again we would outturn them. They used their flaps quite frequently during the fight. One A-36 was on the tail of an ME when the German pilot dropped his flaps and the A-36 nearly crashed into him. A total of five ME-109s were destroyed, with one probable and three damaged.

Liri Valley, 1943

Told by Roy B. Broster, Jr.
Captain, U.S.A.A.F.

The mission for November 2, 1943, was to bomb a grade crossing east-southeast of Rome in the Liri Valley. Twelve A-36s (early model P-51 Mustangs) took off at 1400 and were over the target at 1500. Weather was overcast from 6000 to 8000 ft over the target. Flak was of light and heavy caliber and was accurate. Two airdromes in the area had nothing on them. A train was in the station at Sora with steam up, pulling six cars. Fifteen motor transports were seen going north.

We dived through a hole in the overcast and bombed a bridge over the Liri River. I was flying wing on Tom E. Doyle who had a bomb hang under one wing. We were three and four in red flight. The squadron rendezvoused southeast of the bombing site. Tom pulled back up through the overcast to try and shake the bomb off, but it stayed with him. We turned back to the west toward the coast and, while he was trying to shake the bomb, we had spread apart so he would have maneuvering room. We were crossing the last ridge of mountains before being over the Pontine marshes on the coast when we were jumped by nine 109s. I pressed the mike button to call them out to Tom as two were starting a diving pass on him.

As I pushed the button, I started getting hit with 20-mm and machine-gun fire. I looked around and saw a 109 right on my tail. Doyle was on my right, so I kicked the nose around toward him as if I was closing up on him, crossed controls, and fell off on my left wing. I watched as the 109's tracers moved off to my right as he tried to get a deflection shot at me. I kicked the 36 around into a straight dive and pushed the throttle to the stop. Doyle's radio was not working and he did not hear my call, but he saw the 109s about the same time.

By then, the bomb had fallen away, so he started to dive also. We dove down the side of the mountains and hit the deck over the marshes. We pulled away from the 109s and outran them to the deck, but they continued to chase us for 15 to 20 mi. When they finally broke off, we turned east-southeast parallel to the coast, which was out of sight. We remained on the deck. Doyle flew by dead reckoning until he thought we were off Salerno, where he turned to the coast. Shortly, the Isle of Capri rose out of the sea. We climbed to about 5000 ft and proceeded to our base at Cappacio.

I could see on the way home that my left wing had been damaged by a 20 mm and I could see a few more bullet holes here and there. Upon landing, however, the crew chief asked me if I had seen the tail. I had not because the 36 was blind to the rear unless you went through some contortions to twist around. When I got out of the airplane, I found that a large portion of the rudder was gone. Incidentally, I was not flying my plane that day. I also learned that my coolant line had been creased. Doyle had his oil line nearly severed, and his crew chief told him his engine would only have run 1 or 2 minutes more.

Last Mission for Lieutenant Hudgins

Told by Ernest C. Perry
Captain, U.S.A.A.F.

This mission took place in Italy in the latter part of January 1944. The Germans had conducted a withdrawal and managed to hold the Allied advance to a snail's pace. It ground to a complete halt in January 1944, after 4 months of costly fighting. The Allies were stopped by the heavily fortified Gustav Line, which the Germans had built in the rugged terrain of central Italy in the Cassino area.

Our missions were in support of the U.S. Sixth Corps and the U.S. Fifth Army. At this time, there was a major offensive taking place by the Allies to break through the Gustav Line. We were briefed on our mission, which was to make a photo run at Cassino, recon the highway and railroad system to Rome, and return by way of the Liri Valley. We were to check the roads and railroads for damage done the afternoon and night before. Lieutenant Hudgins was to make a low-level photo run at Cassino. The British Beaufighters and A-20s had been attacking supply lines that night between Rome and Cassino. The U.S. fighter bombers had been destroying highway and railroad bridges. We were to check the bridges to see if repairs were underway. The week before, U.S. fighter bombers had done extensive damage to the railroad marshaling yards between Rome and Cassino. Hopefully, this would hinder supplies from getting to the front.

This was my 10th mission and Lieutenant Hudgins had 40 plus. We took off just before daylight. The Germans did most of their movement at night because the U.S. fighter bombers had been very successful at destroying most anything that moved during the day. We had been catching their movements on these first light missions before they hid for the day. The locations were noted and passed on to the fighter bombers.

As we approached Cassino, you could always spot the front by the heavy barrage of enemy flak. As we neared Cassino, we were flying above haze layer. I spotted two bogies at three o'clock in a vertical dive; they appeared to be FW-190s. I called them out to Lieutenant Hudgins as they disappeared in the haze below. At this time, Lieutenant Hudgins started his dive to pick up speed for his photo run at the front lines. I was on his right wing flying line abreast when I spotted four Spitfires diving toward us from the nine o'clock position. I called them out and identified them as Spitfires. He was busy lining up for his photo run and did not reply. We were also receiving heavy automatic weapons fire from the mountains on either side. They were above our level and firing down on us. About this time, the Spitfires closed and started firing at my aircraft. All four made a firing pass at me, but their tracers went under my right wing as they disappeared in the haze. I lost contact with Lieutenant Hudgins after turning into the Spitfires and never saw him again. It is uncertain if he was hit by flak or enemy aircraft, but the wreckage of his Mustang was found later on an abandoned German airfield in the valley north of Cassino as the Allies advanced.

I found after returning to base that we had flown right in the middle of a German fighter bomber attack on the Allied lines, and the low cover of Spitfires was tangling with the Germans. As their high cover flight dove down to help, they mistook us for ME-109s. We were flying the early model Mustang (P-51), which looks very similar to the 109. The visibility was less than 2 mi in the haze and smoke, which would also be a factor.

In conclusion I am very thankful the Spitfires were unable to close on my aircraft because of their speed. I flew a total of 80 missions on my tour, but this one is still very vivid in my memory, even after 50 years.

It's a Beach: All That Saltwater and No Sand

Told by Kenneth West
Lieutenant Junior Grade, U.S.N.R.

O ne of my most memorable flights from a carrier isn't even in my logbook. The reason is probably because it lasted only about 10 seconds. It was sometime in mid- to late January 1944, approximately 60 mi south of Bermuda on the U.S.S. *Hornet's* shakedown cruise with Air Group 15 aboard. Two things happened: my first crash and my first swim in saltwater.

I was directed into the takeoff position, given the windup, indicated ready, and cut loose. My forward motion had hardly started when my Hellcat took a hard 45° cut to port. "Mister Automatic." That's the way we were trained. Reaction: 60 ft to liftoff, 60 ft to water. Go for it. A brake job wouldn't work. I released the throttle long enough to unlock the gear before clearing the edge of the deck, pulled the stick to my gut, and would probably have prayed if there had been time. For a brief moment I had it, but a 6- to 8-ft wave reached up, caught the port wing, and wiped out the best opportunity available.

The next 30 seconds taught me a lesson. When they told me the backpack shouldn't be hooked to the seat pack until airborne, someone apparently had already learned the lesson the hard way. I waited for the plane to settle in and hit the seat-belt release. Didn't get it or it was hung up, so I did it the second time. Voilà! Started to climb out, but the seat pack wouldn't come with me, so I sat down and unhooked it just as the Hellcat headed for the deep six.

The rescue can was headed my way, probably before I mushed in. However, I decided a ride in my raft would be appropriate. The first attempt to get in was a disaster due to the waves, but all was comfortable when they came down in the cargo net to haul me aboard.

My only injury, unknown until aboard, was a hole in my arm above the elbow where it had hit the radio button hard in my release of the seat belt. The doc fixed me up with a little libation for my ordeal, and within 30 minutes, a breeches buoy ride took me home, which was a disappointment as the destroyer was headed for Bermuda to pick up mail.

Captain Browning wasn't very happy about losing one of his planes on the initial shakedown. Perfectly understandable. The only thing that saved me was a Curtiss rep aboard, observing the SB2C "beast" in operation. He came to the bridge while Browning was interrogating me and told him the left brake had locked on takeoff. "Nuf said."

It has always been my opinion that I would have been deep sixed had I not unlocked the wheels. As I went over the side, the plane tore the armor plate off two 20-mm guns and broke two cables on the catwalk.

Just my lucky day, I guess. In fact, I'd say one of many.

Opportunity Missed

Told by Beverly W. Landstreet
Captain, U.S. Marine Corps

My squadron was the first Marine all-weather squadron, which meant we flew regardless of the weather, day or night. Our mission was to position single planes every dark hour over various Japanese-held airfields to intercept shipping planes and troop replacement. This tour began in the northern Solomons, which included New Georgia, Bougainville, New Ireland, New Britain, and other enemy-held islands.

Prior to entering combat, our squadron was based at Turtle Bay on Espiritu Santo, the largest island in the New Hebrides group. Our first briefing was conducted by an Australian coast watcher located on Japanese-held Bougainville in the northern Solomon Islands. All coast watchers had lived and worked on coca plantations. The briefing concerned survival in case we were shot down. Just as in New Guinea, these islands were inhabited by practicing cannibals, but there were also some Christian natives along the shore. In case we were shot down on New Britain, we were told to avoid the highlands, which was home to the cannibals, and immediately head for the shore. The shore natives, depending on their mood, had a tendency to turn you over to the Japanese. The coast watcher told us our best bet was to inflate our life raft and go to sea, trusting our luck with the sharks. He also said that he and other coast watchers located on every enemy-held island would try to rescue any downed pilot or radio to "dumbo," the PBY-5A's rescue squadron. Many pilots were saved by these courageous men. As General Vandergrift said, "The coast watchers saved Guadalcanal and Guadalcanal saved Australia."

The Bismarck Archipelago, consisting of New Britain, New Ireland, and the Admiralty Islands, is located northwest of the Solomon Island chain. Rabaul, on the northern tip of New Britain, represented a major harbor for the Japanese fleet and was an aviation staging area. In the spring and summer of 1942, the Japanese, in their southeasterly thrust from the Philippines, had occupied all these islands. They built airstrips and naval bases while stairstepping on their march to Guadalcanal and attempting to invade Australia.

Rabaul was known as the garden spot of the South Pacific prior to World War II. It had a perfect deep water port that was bound on three sides by three mountains formed by volcanoes. After the Japanese occupied the Philippines, Rabaul was next. They proceeded to build five airstrips around this almost perfect port. Rabaul was never recaptured; instead it was bypassed but not neglected. Rabaul was bombed and strafed night and day. The same was true of Kavieng on the northern tip of New Ireland.

Only after the end of the war in the Pacific did we discover that the Japanese had 300 mi of underground passageways and underground caves for submarines. These subs only surfaced when they reached the underground port. We wondered why they never seemed to lack ammunition. Now we know!

This brings me to both an interesting and frustrating event. I had been flying in a stationary front over Rabaul for about an hour just before daylight. I had bombed and strafed them during that time, and since I used up all my ammunition, I headed back to Green Island to refuel. I had been in the middle of turbulence, lightning, and all that goes with a frontal system for a couple of hours. So I let down to about 600 ft over the water. Generally, the bases of these frontal systems were under 1000 ft. There you only dealt with heavy rain and intermittent lightning. Suddenly, I was in the clear. Lo and behold, I was facing a Japanese submarine that had surfaced. There were several sailors on deck. The clearing between thunderheads was only about a half mile in diameter. I had no ammo or bombs, only a couple of flares. The sub immediately began to weave as I made a pass dropping a flare. Believe it or not, the sub dove with three or four sailors still on deck. I circled again and dropped the other flare. I radioed the approximate position of the sub, with the Japanese sailors still floating in the Pacific Ocean.

After landing at Green Island, the debriefing officer determined that the sub was probably either waiting to sub-surface at Rabaul or had already unloaded its cargo. I had noticed that there was no deck gun, so we surmised it was a supply sub bringing or leaving ammunition, food, troops, and other supplies to Rabaul. What an opportunity missed! That sub would have been my opportunity to claim a sub kill, an easy one at that.

Listen and You Might Get the Message

Told by Leslie E. Traughber
Captain, U.S.A.A.F.

The pilots in the class of 43-G graduated from Eagle Pass Texas Advanced Flying School on July 27, 1943, and received their commissions as second lieutenants at that time. About 35 of us were sent to Dale Mabry Field in Tallahassee, Florida, after a 10-day leave of absence. Approximately 27 of that group were sent to Hillsboro Field in Tampa, Florida, to begin combat fighter training.

After completing fighter training in P-51As and A-36s, which were P-51As modified with dive breaks, this group of 27 pilots was sent to the port of embarkation (POE) at Newport News, Virginia, and departed on the ship *The Empress of Scotland* to sail unescorted for North Africa. This ship was large: It had three smokestacks and was the largest ship I had ever seen. Of course, I had never seen very much of anything since I was a small-town boy from Jackson, Tennessee.

As fighter pilots, we were supposed to have the better eyesight so we were assigned to submarine watch during the approximately 5-day voyage to Casablanca, Morocco. It was on Christmas Day that we entered the harbor of Casablanca and were given Christmas dinner before we disembarked for an old French Foreign Legion camp. We stayed there for several days before boarding a train for our trip across North Africa. This was in December of 1943, and was it cold on that 40&8 boxcar we traveled in. A 40&8 was a boxcar used in World War I designed to carry 40 men or 8 horses. With 27 bodies and enough K rations for a 5-day trip across the continent to Constantine, there was very little room for all of us to lie down and sleep at the same time. We stood most of the time peering out the side doors at the countryside. Some rode on top of the train and fired their .45-caliber pistols at any target they could find. To add to our discomfort, many of us had diarrhea, and there were too frequent stops and none were timely enough.

We were all a mess when we arrived at Constantine, and a shower in the open air in the dead of winter was a welcome experience. After cleaning up and getting some better food, our problems cleared up in the few days we were there. Within 3 days, we were sent to join the 27th Fighter Bomber Group at a place called Guardo, which was very close to the invasion site of Salerno. They were in dire need of replacements, and we were a welcome sight I am sure.

There were two fighter bombers groups in the area, the 27th and the 86th, flying A-36 fighter planes, and we were happy to be joining a group that was flying in the plane we were trained in. During the advance of our forces from Africa to Sicily to Salerno, the group's losses had been heavy in pilots and planes. In a short time, there were too few planes to equip both groups, and there would be no replacements, so by the toss of a coin, the 86th won and got to keep the A-36s, while the 27th awaited the arrival of the promised new P-51. The 27th had to return to service some old P-40s that had already seen their better days. We never got the P-51s. We later learned they were diverted to England, and we received the new P-47 Thunderbolt, the "jug" as it was so affectionately referred to. This turned out to be the better plane for the type of work we were doing anyway. It could take a lot more punishment than the inline liquid-cooled engine. I later wished I had been in one.

After arriving on January 14, we were given several days of orientation flights, which helped us learn how to use the dive brakes and get the feel of flying again. We were then ready to begin our first combat mission. We not only had to get used to being shot at, but we also had to get over the concern of using the dive brakes, which we were not allowed to use in training because they were too dangerous. Well, we got over the concerns of the dive brakes but never at being shot at.

It was on my 35th combat mission after being assigned to the 522d Fighter Squadron, now flying the P-40s, that I nearly became a casualty of this war. I was shot down and found it necessary to become a member of the Caterpillar Club. This is a distinguished group of pilots who had their lives saved by the use of a parachute. It was on March 16, 1944, that our squadron was assigned a 12-ship mission to bomb an infantry sight at Fontana Lira in the vicinity of Cassino. On this particular day, we probably experienced the worst turbulence I have ever flown in. Up- and downdrafts were so severe that our altitude changes were from 1000 to 2000 ft. This made us a very difficult target from the ground, but it was very hard to fly formation.

Our flights were designated as red, white, and blue flights for communication purposes. I was a wingman on the squadron leader's wing in red flight. As we approached the target area, red leader rocked his wings, which was our signal to get in line of stem (flying one behind the other) and to deploy our dive brakes. We were all over the sky that day, and due to the turbulence, it was difficult to recognize any kind of formation we were flying. Somehow, we all arrived over the target, and red leader rolled over on his back, did a split-S, and dove toward the troop concentration. Each plane is then supposed to reach the same spot and follow him down the same flight path. The dive brakes when deployed slowed our descent, reducing our airspeed and allowing us more time to locate and effectively hit the intended target. With time and experience, we were becoming pretty good at it. We needed to because we were operating very close to our troops on many occasions.

Small arms fire over the target was reported to be intense that day, and I was hit while in my dive. It only takes one hole in your radiator to knock your engine out. If you have ever experienced a leak in your automobile radiator, you can realize you don't have much time before the engine overheats and ceases to function. That is what happened to me, so I immediately attempted to gain back as much altitude as I could before my engine stopped. I also wanted to get out of enemy territory before I had to decide how to solve my predicament. We were flying over mountainous terrain, and there seemed to be no clear area to crash land my plane. With the high winds, bailing out wasn't a good choice, but it was the only one left at the time. I called radar control to establish contact and to try to make sure that when I did bail out I was in friendly territory. While all of this was going on, my engine had stopped at about 10,000 ft and I had been gliding closer to friendlier territory. When I got down to about 2000 ft, I had to do something. Of course, this was 2000 ft above sea level on my altimeter and it was in the mountains, which makes the surface much closer. I don't know how low I was when I got out of the plane, but it wasn't very high above the ground. I made the decision to bail out when I saw a red cross on top of a building below me. I thought this is the place for me, possibly a hospital. I later found out it was in a small town called Senna Aurunca.

When I started to leave the plane, it took several unsuccessful attempts to get out. I climbed in and out of the cockpit twice. My parachute kept hanging on the canopy, which I had neglected to jettison, but you don't always think of everything in conditions like this. Besides, we never get to practice it and most people who had done it can't remember how they got out.

When I cleared the cockpit on the right-hand side, I can remember seeing the vertical tail coming toward me, and as I was about to pass over the horizontal tail, I rotated my body and struck my left shoulder on the vertical tail. This jolted my Ray Ban glasses off, which had been very hard for me to get and I surely hated to lose them. They were suspended in front of my face for a moment, and as I reached for them, I simultaneously pulled the rip cord to deploy my parachute. I suddenly slowed but my glasses didn't.

Since I was so low when I cleared the plane, I made no more than one swing in the chute before I struck a fork in a tree which helped break my lateral speed caused by the high wind. I sheared that part of the tree off, which I believe cushioned my fall, and I only experienced a bruised ankle. No Purple Heart for that.

Before I could get up off the ground and gather up my parachute, I was greeted by friendly faces. What a relief! I was in an orchard used by the British as a motor pool. I remained with them for several hours, and since it was near noon, they provided me with lunch, though I can't remember what I had or if I could even eat at that time. I was told that the mailman would be by soon and I could get a ride back to Naples with him. I then caught a plane back to my airfield at Pomigliano. I arrived just in time for supper and the nightly movie.

Since I was allowed to keep my parachute, I had sheets and a pair of pajamas made from it and maybe a scarf or two. The sheets made sleeping in an army cot really great, and sleeping in silk was a real luxury.

One last bit of truism. We pilots usually had a routine we rarely varied from, and most of us believed there was a higher power looking after us. So this is my story. In most of the places we were in, we were able to get the local people to do our laundry, but on Corsica, the Catholic nuns helped us out. In a very subtle way, I got the message because every time my pajamas were brought back from being laundered, they returned with the fly sewed up.

After 114 combat missions, I returned to the United States on September 16, 1944. I became a P-51 combat instructor at Perry Army Air Base at Perry, Florida, and was later discharged from the service in September of 1945.

The 64th Navy

Told by Edward F. Jones
Captain, U.S.A.A.F.

A light plane, probably an L-5, crash landed on our strip and was abandoned by the Army, its rightful owner. The main damage was to the landing gear, and in the evenings at the officers' mess, we decided there must be some way to make a seaplane out of the little puddle jumper. With the help of talented and cooperative ground crewmen, a set of pontoons was fashioned from sheet metal, the engine was tuned up, and we trucked the finished job to a quiet little lagoon near our quarters. Unskilled in the ways of seaplanes, we had considerable trouble getting it into the air. As the pontoons picked up speed, they formed a vacuum between the bottom of the pontoon and the surface of the calm water. With enough rocking, you could usually get the plane in the air after an extended run. But we had a better way!

We liberated a speedboat that had been abandoned by some Italian summer citizen, got it tuned up and running, and then drew straws to see who would drive the speedboat in front of the speeding seaplane to stir up enough wake to jump the plane out of the water. It worked almost every time.

European Theater of Operations (Italy): Combat Mission 60

Told by Herman K. Freeman
Lieutenant Colonel, U.S.A.A.F.

The following mission will always stand out in my memory. It was the second mission I flew that day, February 19, 1944. The first was a normal dive-bomb mission and then patrol of the beachhead area at Anzio. Nothing out of the ordinary happened. The second mission was basically the same as the first. It was a flight of four P-40s, each carrying a 1000-lb demolition bomb under the belly and six 20-lb fragmentation bombs in the wing racks. The mission was to hit targets of opportunity in the area around the beachhead and then patrol the area for the remainder of the time. There was no obvious movement of troops or equipment for us to attack, so we concentrated on cutting the German supply lines by bombing a road crossing a ravine several miles east of the German lines. After that, we started our patrol of the invasion area. The purpose was to keep any German aircraft from entering the area to bomb and strafe the Allied troops.

While on patrol, we observed one of our heavy bombers, a B-24, circling the beachhead at about 2000 ft altitude. Every time the plane would be over the area furthest from the front lines, a couple of the crew would bail out. We figured the plane must be very badly damaged to abandon it like that, and after the entire crew was out of it, the plane flew directly east into a mountain in German territory and exploded. We heard later that it had been hit by antiaircraft fire, fuel lines had been cut, and it was a flying bomb that was too dangerous to try to land.

Shortly after that, we got a message from our ground control that there were two German aircraft on the deck over Anzio heading north. We were at about 5000 ft and we looked at the area where they were reported and saw them. My flight leader immediately rolled over and headed down to intercept them. I dropped back a little to cover his tail while he attacked. He was in about a 20° dive getting behind one of them, a Focke-Wulf 190, just as he went across the beach heading out over the sea. He gave him a couple of short bursts from his guns, and I could see the bullets hit the water behind the German. As soon as he realized he was being shot at, he made a hard right-hand turn and headed inland. When he did that, my flight leader pulled up and to the left and let him go. I knew he was going to get away, as I knew he hadn't been hit, so I called on the radio, "I got him. Cover me!" I cut him off and dropped in on his tail about 75 to 100 yd behind him. I started to fire, and after several good bursts, he began to smoke very badly. We were right on the deck and it was really rough flying, as I was in his prop wash, firing at him, and wondering if he was ever going to crash. I knew he couldn't get far by the amount he was smoking, but I wanted to see him blow up or crash. I then heard my flight leader call me to join up with the flight over Anzio, so I broke off the attack.

As I turned right and pulled up, I felt a terrific jolt and saw a big flash of fire. I was only 15 or 20 ft off the ground and I didn't know exactly what had happened, but I had to hold the stick clear over to the right to keep the plane from rolling violently to the left. I checked my left wing and there was a gaping hole in it just about where the insignia was supposed to be. The engine was running well and I could control the aircraft okay even if it took all my strength. I decided I would stay with it as long as I could. We had about 90 mi over water to fly to get back to our base and I was plenty beat by the time we landed.

I was really elated that I had gotten my first victory, and when I got into operations to report it to the intelligence officer, I was told that my flight leader had already claimed it. I told them my story about chasing him clear across the beachhead and using up most of my ammo, so someone suggested that we flip a coin for it. I said I'd rather do that than have half a victory and my flight leader agreed. The intelligence officer flipped the coin, said, "Call it, Herm," caught it, and put it on the back of his other hand. I called heads and it was heads. My flight leader said, "That's no way to flip a coin. You got to let it drop." I was getting pretty tired of this, so I said to go ahead and flip it again. He flipped it again, I called heads again while it was still in the air, and when it came to rest it was heads again. I finally got credit for the kill, but the fight over Anzio was easier than the one I had to win on the ground.

Later, when the official reports came in from the front, it was officially confirmed that I was the one that shot him down. The Focke-Wulf crashed moments after I turned off, and the location of the crash agrees with my story. Also, the amount of ammunition expended in our aircraft was checked. My ammo boxes were almost empty and my flight leader's were almost full. There was enough hard evidence to give me the victory.

The hole in my left wing was checked and it could not be determined if it was a 40 mm or 88 mm that hit me. The entry hole in the bottom of my wing was about 8 in in diameter and the exit hole on top was about 2 ft in diameter. I sincerely doubt it was an 88 mm. I don't think I would be telling you this story if it were. Whatever it was, I'm lucky to be here.

One other thing you may have noticed in this narration. I never mentioned my flight leader by name, as he is still alive and well and is still perturbed with me because I stood up to him and claimed my victory. He was one of the old heads and a captain, and I was a new head and still a second lieutenant!

Red Haze

Told by William S. Miller
Captain, U.S.A.A.F.

On a typical, hot, cloudless afternoon in Karachi, at the huge Malair cantonment, Bill, Roggy, and Bob were bicycling over to the airstrip. Last night, W. O. Moon, the squadron engineering officer, had told them that at least one of the P-40s they were to get would be ready for a test hop today.

Their 81st Fighter Group, for reasons known only to God and the War Department, had been transferred from Naples, Italy, to China. They were waiting in India for their new P-47s to come in by Jeep carrier. In the meantime, they were to get several loaner P-40s for the squadron pilots to keep their "hand in." Bill, Roggy, and Bob had all flown the P-40 Warhawk last year in Panama, so they were to test hop the new birds and were to act as instructor pilots for the other 20 to 25 pilots who had never flown the plane. The squadron had previously been equipped with the P-39 Aircobra in Italy. The big 12-cylinder Allison engine powered both planes, so the other pilots would know how to operate it. However, after flying the sweet ground handling, tricycle geared P-39, the clumsy tail dragger P-40 was going to give the new guys fits.

Although these three pilots had all been stationed in Panama, they had first met at a reprocessing center at Dale Mabry, Florida. They then rode together in a troop ship to Africa, endured a 5-day train ride across North Africa, and had arrived in Italy on Christmas Day 1943. They had become fast friends, roommates in a cold tent in so-called sunny Naples, and were all 19 years old. The old guy in the flight was almost 21.

The three friends were excellent-to-superior pilots. One of them had once coined the phrase "We are not as good as they say we are; but, we are pretty damn good!" The 1–10 scale had not been invented at that time, but if it had, Bob would have rated an 8.8, Roggy a 9.3, and Bill a 9.5. Bob lacked only a strong killer instinct which the other two had. Roggy was fated to be a permanent wingman; Bill was blessed or cursed with the look and feel of a leader. Since their arrival in the squadron 4 months ago, Bob had worked up to 19–20 on the pilots' pecking order, Roggy was about 17–18, and Bill was up to 12–13.

They were forced to bicycle to the airstrip because unlike Italy, where each flight had had at least two Jeeps, transportation was almost nonexistent. All three were dressed in new short-sleeved khaki shirts replete with China, Burma, India, and 14th Air Force patches and embroidered rank insignia and silver wings. They wore overseas caps with their appropriate insignia. Bob and Bill were second lieutenants, and Roggy was a flight officer. Of course, they each wore sunglasses and each was shod in a brand new pair of handmade 11-in slip-on boots. The boots cost around $10 a pair, and most of the squadron pilots had two or more pair.

When they arrived, Moon told them that one of the P-40s would soon be ready to test hop. Since Bill had a few more hours of P-40 time and they all knew he was the best pilot, he won the first flight honors with no argument. He asked Moon to scrounge up a Jeep or weapons carrier to run him back over to the BOQ to get his helmet, parachute, flight suit, and GI shoes. Moon checked the flight line and the squadron area and found all of the limited vehicles were on trips, tied up, or missing. It was not physically feasible to walk or bike 7 miles in the 100-degree-plus heat with a heavy parachute and the rest of a pilot's flight gear. Besides, it would have wasted approximately 1.5 hours of flying time.

Accordingly, Bill asked Moon to get the plane ready as soon as possible, and he borrowed a throat mike, a set of earphones, and a loaner parachute from nearby base operations. A throat mike consisted of an adjustable elastic band, GI olive drab, of course, necessary wiring, and two soda bottle-cap-sized sensors which fit tightly on each side of a pilot's Adam's apple.

These P-40-N had been in combat in China and the India-Burma theaters and had subsequently been assigned for use in that theater's fighter indoctrination school at Karachi. This ubiquitous "new pilot" school gave them an orientation check-out in the aircraft and some rudimentary precombat training. Each aircraft had over 800 flying hours and they were *real war wearies*. The 91st pilots were to find that they flew something like a used farm pickup with 120,000 plus miles on them. The engines were okay technically, as were the airframes. However, the wings, fuselage, ailerons, rudder, elevators, and flight control levers were loose as a goose.

In view of the above, Bill gave this bird an extra special pre-flight inspection as he looked into all the nooks and crannies for leaks, loose nuts, chaffed or frayed fuel and oil lines, and so forth. Finally, by about 1415, he was ready to try to get the "bucket of bolts" into the air. While taxiing out to runway 90 for an easterly takeoff, he had the same surprising sensation he had noticed on his first P-40 flight: What a huge, long nose the aircraft had even though it was a fairly small fighter. Especially with the bomber-transport earphones instead of a helmet to help muffle the sound, the engine noise was deafening. To see where he was going, it was necessary to taxi in sweeping S-turns instead of going straight forward. This required a lot of brake stomping since the P-40 was cursed with a notoriously poor set of brakes.

Of course, the engine overheated and the coolant tempera-ture for the liquid-cooled Allison engine went into the red. Never mind, this was normal in hot tropical or desert coun-tries and enough of the 15.5 gal of coolant always seemed to be left to get the job done. The takeoff preflight went okay, so he got tower clearance, rolled onto the runway, and gave it the needle—full forward on the throttle.

He was relieved to feel that all the flight controls evoked a proper response, the airspeed indicator worked, and he was getting almost full-rated power out of the engine. Instead of 44 in of manifold pressure or boost, as the British called it, he was getting 40 and the propeller revolutions per minute were 2700—very close to the prescribed 2800. The fuel oil and hydraulic pressures were in the green, and the cylinder head and coolant temperatures were falling toward the green lines.

Bill remembered and was prepared for the strong torque which tried to pull the aircraft off to the left of the runway while cheerfully trying to drag the left wing tip on the ground. But it was still a surprise to feel its strength. What was then called "torque" is now called the "P factor," and it is caused by the relatively tiny propeller, which turns to the right, actually trying to roll the much larger plane to the left. Again, he noticed the deafening engine and slipstream noises as they approached takeoff speed of 70 mi/h. "Wish I had my nice British helmet," he thought.

Soon, the plane lumbered into the air, and Bill retracted the gear. The aircraft yawed several times to the left and the right for Curtiss Wright never did learn to make the gear come up simultaneously. He milked up the flaps in several small increments to reduce the "sinking elevator" feeling as wing lift decreased. By then, he was at around 50 ft and was climbing at 2000 ft/min.

He noticed it was a beautiful afternoon. There was the faintest suggestion of a haze layer at 5000 ft, but even from there, the visibility was at least 20 to 25 mi. There were a few scattered, fluffy cumulus clouds out over the coastline. The sea across from Karachi's protecting sand spit was a beautiful deep blue and was almost as pretty as the sky above. He thought, "If we can scrounge up a Jeep, we can drive into Karachi, find the boatman, and sail over to the sand spit for a swim later this afternoon."

He leveled the P-40 at 6000 ft and started his normal procedure when he checked out a new airplane or when flying one that he hadn't flown recently. This included a group of stalls in differing configurations, several simulated landings, spinning the bird, using it as a weapons system, and finishing with aerobatics.

He did several clean and dirty stalls, with gear and flaps up and then down, in various flight positions. These were all normal with no unusual snaps, pops, or cracks. Then he picked out a simulated landing altitude of 3000 ft and did several practice landings. Each time, Bill noticed that he always ended up in the minus category at about 300 ft below the simulated runway altitude. This had concerned him considerably last year when he first flew the P-40, and he had approached the ground very gingerly. In those days, they didn't know about the phenomenon called "ground effect." This effect occurs as an airplane nears the ground and is really, in simple terms, a packing of the air molecules between the aircraft wing and the ground. This packing results in the wing developing more than normal lift, and it gets progressively stronger as the plane gets lower. Pilots knew instinctively that something was helping them land, but they didn't have the educational background to understand the molecular theory of matter. At least, they were never told about it, probably because the instructors didn't know either.

Next came spins. Spins in almost all the fighters Bill had flown were either prohibited or not recommended. However, Bill and most of the better pilots constantly flew their birds on the edge, or just past it, to get the job done. Bill was not about to try for max performance unless he knew what a spin felt like and knew how to get out of one. This had already come in handy and probably saved his life twice.

After the conventional spins, he was ready to try out the weapons. It was 1440. He turned on the gun sight, climbed back to 5000 ft, and found some wispy clouds which had drifted in from the coast. He shot down several of these "cloud Zeros." After he became an instant ace by shooting down five, he climbed to 10,000 and tried dive-bombing and then dove on down to the deck near ground level and strafed numerous sand dune "tanks, trucks bunkers, etc." It was then 1450.

Bill was beginning to feel pretty good about the old "bucket of bolts." The P-40 was a miserable airplane by anyone's standards, and this particular plane was one of the worst of the species. Nevertheless, flying it was much better than not flying at all. He mused that the only thing really wrong with this airplane was that it had been years since anyone had loved and cherished it. It had seldom, if ever, had a bright-eyed young pilot or a loving crew chief to polish it, talk to it, or care for it.

He had about 10 more minutes to play before going in to shoot several real landings. It was time now for the icing-on-the-cake aerobatics. He climbed up to 5000 again and did a batch of slow rolls, barrel rolls, and other rolling maneuvers. Then came several loops. In order to do loops in the P-40, one was supposed to get about 350 mi/h before entering them. They could be done much slower, but they got sloppier the lower the entry speed was. Satisfied with his loops, he then went into an Immelmann. The Immelmann was a half loop followed by a half roll while inverted on top of the maneuver. For this maneuver, 370 was the normal recommended entry speed.

As Bill started his 4-G pull-up at 370 mi/h from 5500 ft, he sensed a flutter somewhere in the upper left of the cockpit. He ducked forward instinctively as rapidly as was possible without hitting the gun sight, which was only 5 in away from the pilot's face. This was the last thing he remembered until he awoke in a red haze.

He came to consciousness in a gentle dive at 300 mi/h and passing through 3000 ft. He could scarcely see the instruments due to the thick red fluid in his eyes and all over his forehead. On tasting the fluid, he realized that it was blood rather than hydraulic fluid. He also became aware that something on the left side of his head was flapping in the breeze. He tried to feel what it was, going on with his right hand, but found it would only go to shoulder level. As his battered brain slowly awakened and started computing, he realized that the canopy was missing. The canopy on the P-40 was a U-shaped affair of aluminum and glass which slid on a track so the pilot could enter, leave, and windproof the cockpit. It was roughly 32 in long and 30 in tall and was built like, and often was called, a "greenhouse." It weighed maybe 60 lb and had many nice jagged edges, the better to cut you with. Apparently, the left front roller had sheared, and the whole mass had dipped into the cockpit from left to right at 350 mi/h. It had either taken a sizable piece of his skull off, had scalped him, or both.

Bill gingerly felt toward his bleeding head with his left hand and found that his left side was working okay. As he cautiously explored the damage, he realized it was a section of his scalp near the size of half a football that was flapping. The canopy had struck him on the left temple just where the skull starts rounding backwards and had gouged a path on the top of his head. He thought, "If I hadn't ducked, it would have taken off the top of my head."

What to do? Bill reduced power to 20 in, jacked the seat to the floor, and trimmed the bird to hold 180 mi/h. Next, he tried to staunch the blood so he could see to fly. First, he aimed the plane westward and tried to make a turban with his handkerchief. It was slightly too small and, of course, promptly blew away. His next idea was to take off his shirt and try to wrap it around his head turban style. This required him to unlatch his shoulder harness, open his seat belt, get out of his parachute, remove his shirt, and wrap it around his head. As he tried to do this with his semiparalyzed right side, it soon became evident that it was going to be very difficult to accomplish. He realized he didn't have the strength or the patience to do it.

Bill then experimented with holding his scalp down tightly with his left hand as he rested from the attempt to remove his shirt. He found that this made a considerable difference in both the flow of blood and the discomfort from the flapping scalp. By swabbing his eyes with his left fist, he could see well enough to head the airplane in the right direction.

At this stage, Bill decided that it would be very appropriate to call for help. Though not a lapel grabber, he was a confirmed Christian and believed in the power of prayer, so he first asked the Lord for help. He said, "God, I don't think I can make this one by myself. If you want me to make it, and

I hope you do, I am going to need some help." Next he thought about trying to contact the control tower for help. However, he realized for the first time that his throat mike and earphones had been blown off.

The throttle mike button was wired hot and made a squall when pressed, so he decided to send the tower an SOS in Morse code. He transmitted "SOS P-40" several times. He would have liked to give them his tail number, but he couldn't remember the number and couldn't read the cockpit placard that showed it.

Like mushrooms popping in a wet yard, the thought, "If I only had my helmet," kept occurring. The heavy, double layered, khaki twill, British helmet with its suede headband would have cushioned the blow. The headband fit exactly where the canopy had dug in. It would not only have lessened the size of the tear in his head but would also have lessened the flow of blood. The goggles on the helmet would have also been valuable in controlling the flow of blood, and he would still have had a way to contact the tower.

As he drifted westward, he gradually became aware that he was passing out frequently and that the duration of the blackouts was becoming longer. His enfeebled brain came to the decision to bail out. Standard ejection procedure in the P-40 was to stand up on the seat and do a belly buster onto the right wing. This gave one a slightly better than 50-percent chance of missing the horizontal stabilizer without losing too many parts. Bill rejected this one because he had neither the strength to stand or jump out on the wing. The alternate procedure was to roll the aircraft halfway with its belly up and then kick the stick forward. This popped one out rather smartly, they said, and again gave one a fighting chance to miss the tail assembly. "Sounds like a winner," he thought.

He checked to see if the seat belt and shoulder harness were unlocked and started a roll to the right. Just as the wings reached vertical, he realized this was a bad idea. The impact of popping out of the plane and the shock of the parachute opening might well pop the rest of his head arteries open. Also, if he missed the tail, he would float down into the barren desert. If no one had heard his SOS, then he would surely bleed to death before they could find him. His fancy boots would also pop off his feet, and he would be barefoot in the scorching sand.

He struggled back into his seat belt, rolled level, and tried to get the harness back on. He was never able to get it over his right shoulder. The effort exhausted him and for the first time he became discouraged. "I may not make it," he thought. He was noticing that, although the red haze was getting thinner, the sky and cockpit were gradually getting darker. "At least the bleeding is slowing down," he thought. "Wonder how much blood I have lost."

He was getting so very tired. "Maybe it would be easier to let go and to go on in," he thought. "It would be peaceful anyway." Fortunately, the pain had not started yet or he would have been very tempted by Death's whispered call, "Turn loose! Let go! It's peaceful over here." Bill appealed to God again and turned it over to him, "God, I can't do it by myself. If you want me to live, you have the airplane." He even waggled the stick as one traditionally does when he turns the controls over to another pilot.

The mushroom thought, "If I only had my helmet," kept occurring. Bill rejected it as he thought, "I have enough problems to solve without worrying about spilled milk, or blood." He droned on toward the base and, through the dark haze, was able to spot the runway. He started a descent into an approach to the takeoff runway, 90° to the east. In the interest of saving time, he would land straight in rather than peeling off in the normal fighter pattern. He slowed to 150 mi/h and went through the ridiculously complicated procedure of lowering the landing gear and assuring it was down and locked.

He blipped "SOS P-40" several more times to get the tower's attention. Slowing to 120 mi/h, he lowered the flaps, checked gas on Main Tank, mixture Auto Rich, propeller Full Increase RPM. As he passed 1500 ft, he lined up toward the runway. He awoke at 1100 ft off to the right of the runway by 200 or 300 ft. "The monster is determined to do me in," he thought.

As he put gear and flaps down and made the necessary left-hand adjustments to trim tabs, power setting, and miscellaneous valves and levers, he noticed that the blood flow was controllable with only an occasional wipe instead of constant swabbing, even when he released the pressure on his scalp. "My blood is about all gone," he thought.

Continuing on in at 500 ft, he checked the gear position markers, noted that the hydraulic pressure was 1000 lb/in^2, and temporarily chopped the throttle to listen for the gear warning horn. He made a final flaps full down check, checked again that mixture control was in Auto Rich, and prop control was at Full Increase. Throttle was at 17 in and airspeed at 90 mi/h, which was giving him a sink rate of 600 ft/min. Bill had never done a straight-in approach in a P-40 before, so this one had to be just like the book said. He noted that it had been about 8 minutes since he was scalped, and the time was 1459.

He came to next at 150 ft way off to the right. He made a steep left turn back to runway heading and checked the runway. He noticed two P-40s in formation were starting to roll down the runway headed in his direction. He thought, "I know I won't last around the pattern again for another try. If they don't see me, then there is going to be an awful mess in the middle of the runway. God help us all!"

The next time he regained consciousness, he was 30 ft off the right side of the runway. The two formation birds were still rolling toward him. He quickly checked airspeed and saw 85 mi/h. Bill eased the stick back into his lap to raise the nose and flare the glide. He eased the throttle forward to 22 in to decrease rate of descent and stomped hard left rudder to get the big nose over to the runway, while feeding in hard right stick to keep the wings level. The aircraft was in danger of doing a snap roll to the left and augering in upside down. This form of desperation and newly invented maneuver worked because just as the nose drifted near the center of the runway, he kicked right rudder to stop the drift, and the aircraft went into a gentle stall about 3 in above the runway. He had literally willed the aircraft onto the runway. The touchdown was surprisingly smooth. "As a matter of fact, I have made worse," he said to himself.

He snatched the throttle all the way back, raised the flaps, and immediately checked for the formation of P-40s. Much to his relief, he saw them leaving the runway to the left and to the right about 2000 ft away. "Bet they are cursing me to a fare-thee-well," he thought. At 2500 ft down the runway, he wished for a feeder taxi strip so he could get over to the tower and the ambulance as soon as possible. He didn't relish having to taxi this big klunker all the way to the runway's end and then back to the tower. Bill considered trying to taxi across the sand but rejected that since he might break a landing gear and nose over in the soft sand.

He stopped his gentle braking and let the aircraft keep rolling toward the end at a brisk 25 to 30 mi/h. With no warning as he neared the end of the runway, the P-40 decided to get even for the 800 hours of abuse it had taken. The left landing gear suddenly collapsed. The airplane did a scraping, graceful 180° turn and the nose dug in. The tail came straight up and the P-40 balanced on the propeller spinner, teetering gently while trying to decide whether to flip all the way over onto its back or to slam down right side up. "Oh, God. Please don't let it flip over," he prayed. "Another lick on my head, which I will get because I couldn't get the other shoulder harness on, will do me in for sure." God heard the prayer and the tail hammered down right side up.

Bill's next concern was fire, so he immediately opened his seat belt, took off the shoulder strap, and unbuckled and shed the parachute while cutting the ignition switch to kill fire to what was left of the engine. He stepped onto the left wing and was relieved to see no sign of fire. He noted that the right landing gear was still okay, but the tail wheel was squashed up into the fuselage. He noticed with relief that a bicycle was approaching rapidly and an ambulance was about half way to him. He thought, "I may as well clean up the cockpit like a good pilot and fill out the aircraft status forms." He cut off the fuel selector, checked the ignition switch off, killed the radio power, and leaned over to get aircraft form No. 1.

"There is a possibility I may never get to tell anyone what happened," he thought. He decided not to log the total flight time of 48 minutes but felt he must make an entry in the aircraft status section. He wrote: "Canopy came off in flight. Left landing gear collapsed on roll out. Aircraft nosed over and bent the prop. Wm. Miller, 2d Lt." He then sat back on the wing and watched the bicycle and ambulance continue to approach. The bicycle won. It was Roggy. Roggy said, "My God, Bill, look at your head! What happened?" Bill asked, "Is it bad, Roggy?" "Yes, it is bad, Bill," Roggy said. The stretcher crew scooped him up at about this time and took him to the dispensary in base operations.

The duty flight surgeon immediately gave him a large shot of morphine and started Novocain shots into his scalp while Roggy asked, "What happened, Bill?" Bill got out, "The canopy came off in an Immelmann," as he felt the first suture of his bleeding arteries. He thought, "The Novocain needle hurt worse than the stitches," and gradually passed out into a deep sleep.

He woke up 3 days later as the resident doctors and the 91st Squadron flight surgeon, Dr. Max Salvater, were making their rounds. Dr. Max's first words were, "Bill, it was touch and go the first night and all day yesterday. We almost lost you. We had to give you three transfusions the first 24 hours, as you were the most dried out living human we have ever seen. The canopy hit you over the temple and gouged its way up your scalp. As it gouged along, it kept getting deeper until at about the center of your head it had to make a decision as to whether to go on into your skull or to bounce off. You will probably have a permanent dent in your skull back there, but you are mighty lucky that it bounced off instead of penetrating your head. You have a fractured skull, a severe concussion, a broken nose with 5 stitches where the right nose piece of your sunglasses went into it, 58 stitches in your scalp, and 7 more above your right lip where glass from the canopy or your sunglasses went through. But the good news is, you are off the critical list and are going to be okay. You surely do have a hard head. Major Chandler and all the operations guys say that you did a fantastic job." Dr. Max fielded several questions such as, "How long? How soon before I can fly again? What are these tubes in my arms?" and so on and told Bill he could have visitors on the next day if he felt like it.

The first visitors were flight leader Ed Weed, Roggy, and Bob Porter. After Bill gave them a full account of what happened, he caught up on their activities that fateful afternoon. Bob had left the flight line shortly after Bill had taken off since Roggy had pulled his "better pilot" rank on him and it would probably have been too late for him to get a third flight that day. Roggy related this story: "About 45 minutes after you took off, I went outside operations and began checking the sky. At about 1458, I heard a P-40 approaching slowly from the east. It was going too slow and it was doing a lot of wobbling so I sensed that something was wrong. The fact that it appeared to be making a straight-in approach instead of your typical pitch up also gave me a clue that something was not normal. I also saw that two P-40s were taxiing onto the runway for takeoff opposite the direction which you were landing. I ran into base ops and ordered the tower to tell the other P-40s to abort because a P-40 in obvious trouble was landing the other way. As I continued to watch those jerky landing contortions, I concluded there must be a bad oil or hydraulic leak, and you couldn't see where you were going. That final landing maneuver was the most God awful thing I have ever seen. Your P-40 was lined up way off to the right of the runway and was wallowing like a pig. At about 10 ft in the air, the nose reared up into a three-point attitude and the plane then began to slide rapidly over to the center line of the runway. Then, abruptly the slide stopped and the aircraft stalled onto the runway in the near perfect three-point landing. I couldn't believe that it didn't crash. As the bird rolled past the tower, I saw the canopy was gone and the side of the fuselage was streaked or covered with oil or some other dark fluid. Then I screamed at base ops to get the ambulance moving and jumped onto my bike. As I pedaled rapidly toward the end of the runway, I saw the left gear collapse and the airplane do the balancing act on its nose. When I got closer, I could see your head and I knew what the dark fluid was!"

When Ed, Bob, and Roggy left, they said, "Oh, by the way, if you were in a uniform, you would be out of it. You made first lieutenant yesterday."

Twenty-seven days afterward, Bill left the hospital, and on the next day, he took another P-40 up on an uneventful flight. He never again flew a fighter aircraft without his own helmet and other flight gear.

Two Combat Tours

Told by J. Pat Maxwell
Major, U.S.A.A.F.

I was a member of the 78th Fighter Group stationed at Duxford Air Base near Cambridge, England, flying P-47 Thunderbolt aircraft. The 78th was one of the first fighter groups to go operational in the Eighth Air Force in the European theater, flying its first combat mission in April 1943. At that time, combat tours for fighter pilots had not been established. Later in the year, tours were set at 200 combat hours, and many of us in the 78th had already exceeded that amount. I wanted to participate in D-Day operations, which were yet to come, so at my request, I was allowed to continue flying missions. Finally, in May of 1944, the Eighth Air Force announced an option to those of us in my category to be returned to our homes in the States for a 30-day rest and recuperation leave, provided we would agree to return to our fighter groups and fly a second combat tour. I agreed and was returned to my home in Pleasant Hill, Alabama. The very first morning at home, my dad awoke me with a radio, no TVs then, blasting the news. The Normandy invasion was underway! It was D-Day, June 6, 1944! I could hardly believe it! I have, however, always believed that my leave was providential. Upon returning to my group, many of my fellow pilots were missing, and there were many new faces.

The 78th Fighter Group played a major role in providing escort to protect Eighth Air Force heavy bombers, B-17s and

B-24s, on daylight bombing operations on targets in Germany and German-occupied countries on the European continent. As the war progressed, however, we were also assigned dive-bombing and strafing missions on targets in advance of our Allied ground forces.

As I neared the completion of my second tour, one particular mission in September 1944 stands out in my memory. This was the Allies' airborne invasion of Arnhem and Nijmegen conducted to overcome the strong German defensive effort as our Allied forces were approaching the German western border. The 78th Fighter Group played a part, with nearly 50 P-47s scouting the assigned troop landing areas and flying at altitudes low enough to draw enemy fire to locate gun positions and then dive-bomb and strafe these gun positions. After destroying the flak positions, we flew back to the Dutch coast and met C-47s towing gliders loaded with airborne troops. We provided escort to the landing areas, where we formed a protective cover in the air above these areas while our Allied forces landed, holding losses to a minimum. For this effort, the 78th Fighter Group was awarded a Distinguished Unit (Presidential) Citation.

In December 1944, the German enemy forces attempted their last major offensive to push through the Allied front lines in the Ardennes known as The Battle of the Bulge. Bad weather limited Allied aircraft from giving full air support; however, the 78th was successful in getting its P-47s in the air to provide escort for our bombers to bomb communication and supply lines. We destroyed a number of German enemy aircraft which contributed to the defeat of this German effort. For this effort, the 78th received special commendation from Lieutenant General Doolittle of the Eighth Air Force. I flew most of these missions and, in fact, led my squadron (the 84th) on a bomber escort mission on my last mission on the last day of the year, December 31, 1944. This was also the last mission flown by the 78th in P-47s before changing to P-51s, and I destroyed a German FW-190 in the air, which was the 78th Fighter Group's 400th Nazi plane destroyed as of that date.

I was 21 years of age when I flew my first combat mission and was extremely fortunate to fly 133 missions, 435 combat hours, and escape injury or being shot down. Now in my seventies more than 50 years later, I am more than ever a believer that survival was to a great extent providential.

Somewhere in Space

Told by Kenneth West
Lieutenant Junior Grade, U.S.N.R.

A s my story begins to unfold, the title comes to me like a bolt out of the blue, "Somewhere in Space I Hang Suspended." I can vividly see the old movie short featuring the comedian singing the story line with his scarf wrapped around his neck like a hangman's noose.

But actually, I wasn't in space and at that time the song was the furthest thing from my mind. I was in a Grumman F6F Hellcat in the Pacific in 1944. Also, as most fighter pilots' stories go, I wasn't flat on my back at 20,000 ft either. In reality, I was in a hammerhead stall at about angels 10, and several things had just happened to get me in this position. (For readers who don't know about angels, it's a Navy term for thousands of feet in the air; the higher you get, the more angels you encounter.) Anyway, there I was hangin' high in a hammerhead and in trouble. Make that *big trouble!* I could use all the angels in the sky right about now. I chopped the throttle and start backing down right into the face of a Japanese fighter headed straight up with guns blazing, I'm boresighted, and I can hear the bullets hitting the canopy. But let's begin at the beginning.

The task group had been at general quarters since before daybreak for we knew that the enemy was in the area. Combat air patrols were in the air and all pilots in the task group were in their ready rooms.

Aboard the U.S.S. *Essex,* in the fighter pilots' ready room, I was one of the pilots on ready alert waiting to scramble at a moment's notice when bogies (enemy aircraft) are spotted on radar and the command, "Pilots, man your planes," comes in over the squawk box. Eight fighters were spotted on the flight deck, their engines warmed and ready to launch to intercept any enemy aircraft. One of those was mine.

While waiting, I checked my backpack and found a .38-caliber ammo pack broken open, so I pulled it to see what could be done with it. At about that time came the command, "Pilots, man your planes." I quickly shoved it into the knee pocket of my coveralls and hit the deck running along with the others to our assigned planes for immediate launching.

Our orders were "Buster," which meant the max climb-out speed for altitude and an intercept for the approaching attack group of about 50 or 60 planes, which were only about 40 mi away. Of the eight planes, one didn't check out. Therefore, there were only seven in the air climbing out for the intercept, with me in the last position.

The flight, normally two divisions, led by Group Commander Dave McCampbell, now consisted of six F6F5s and me in an F6F3; hence, their Buster was much bigger than mine, and as the climb-out continued, they ran off and left me. I don't know if Dave was aware of this or not. I did know that our mission was to stop those planes from reaching the task group, and so I was certainly dispensable because they had to make the intercept before the attack could hit the fleet. In addition, our weather wasn't the best, with some clear sky and a long squall line with clouds building at times to 20,000 ft or more, which could create a problem.

As I continued my climb, it wasn't long until I saw a Japanese fighter sweep by. I swung hard and cut in behind him, caught him on the turn, and opened fire sitting on his tail. The plane exploded and I saw the pilot blown from the cockpit just before I pulled up and pushed the throttle to the firewall to avoid the debris.

Guess where I am now? You're right! So back to the beginning, and there I was...

My first thought was, "I've been had, so take him with me." I moved the controls over hard to clear the hammerhead and noted that most of my left aileron was gone. It was then that I saw .38 shells hanging in the air, so I started catching them and shoving them back in the knee pocket as I backed down and out to meet him. There was nothing else I could do at the moment. It had been my shells, not his, that had been hitting the canopy.

Luckily, he didn't hang around for my greeting and I continued to dive out to cloud cover. There were at least two other Japanese fighters below, but damaged and in no position to attack, I headed for home through the front. Bad choice. After two attempts, which tumbled my gyro horizon, requiring a 180° turn with a needle and ball controlled fly out, I looked for room at the top and found space at about 12,000 ft.

The most fear I ever encountered while in the Pacific was on the flight home. With IFF on, I called that I was heading in, but was not recognized. They vectored out the CAP and it headed my way. Suddenly, dead ahead, four Hellcats were headed straight at me. Can you imagine, bogies all over and the CAP looking for trouble? My heart was in my mouth because I was facing four Hellcats with a total of twenty-four .50-caliber machine guns with what might well be some trigger happy pilots at the controls. All I could do was wait. Thank heavens they didn't fire. They didn't even hang around to escort me home.

Someone caught hell soon after I hit the deck because my plane's right side said "Flyable Dud." It was not to have been flown, except to be replaced. Pilots always enter from the left side. Had I known it prior to takeoff, only six planes would have launched. Oh well, I got home.

Midair Fill-Up

Told by William Shwab
First Lieutenant, U.S.A.A.F.

In April 1944, our 311th Fighter Squadron, 58th Fighter
Group, Fifth Air Force, flying P-47s, was stationed in
Saidor, New Guinea. New Guinea is an immense island just
above Australia, across the Coral Sea, north of Darwin and
Townsville, Australia. All of our operations were along the
northern shore of New Guinea, which is 300 to 400 mi long
from Port Moresby to Hollandia, off the Dutch east island
of Nomfoor. We were stationed on a metal fighter strip in
Saidor, 100 mi north of Port Moresby, headquarters of the
Fifth Air Force fighter command. There was a large
Japanese Army group of infantry and artillery, approximately
40,000 troops, which was heavily fortified at Wewak, 150 mi
north of us. This enemy was fighting our infantry and
cavalry groups.

The weather in this area was extremely hot and humid,
with unusually bad weather occurring mostly in the after-
noons. Most of our missions were dive-bombing missions
with either 500-lb general purpose bombs on each wing or
100-gal napalm bombs on each wing, with a 75-gal belly
tank of fuel for added range. We had a squadron of B-25s on
our strip at Saidor, and on this particular mission, we were
to fly cover for a flight of 12 B-25s over Wewak. Then we
were to proceed to strafe, after the B-25s completed their
low-level parafrag bombing mission. We used the external

fuel tanks until we reached the general target area and then jettisoned them before beginning our strafing runs.

I turned on my gun sight before heading toward the target, and as I began the attack, my engine started sputtering and cutting off and on, and then quit, right on final with my nose pointed down at 40,000 Japanese troops. Also, there was lots of small and intermediate ack-ack all over the place.

I presumed at the time that I had been hit and proceeded to turn out to sea to prepare to ditch. I kept trying to restart my engine and, during this frantic search, found that my fuel switch indicator was on external fuel. I finally turned the switch to internal fuel, hit the booster switch, and the engine returned to life. I went ahead and made my strafing run and returned with my flight to Saidor, our home base. That error about the fuel switch was a great instructor, and that particular situation never happened again.

European Theater of Operations (Italy): Combat Mission 111

Told by Herman K. Freeman
Lieutenant Colonel, U.S.A.A.F.

One mission that stands out very vividly in my memory was my 111th combat mission which I flew on May 18, 1944. The war in Italy was in a stalemated situation along the southern front, which was along the Garigliano River. The front that had been established to the north by the Anzio beachhead was also not able to advance inland due to the heavy concentration of German troops. We were flying many road reconnaissance missions with the primary intent to destroy anything that moved in an attempt to cut all German supply lines. If we didn't find a suitable moving target, we always had an alternate target designated to dive-bomb.

This mission consisted of a 12-ship flight of P-40s, each carrying a 500-lb demolition bomb, twelve 20-lb fragmentation bombs under the belly, and six more frag bombs in the wing racks. This day, the enemy traffic was very light and not worth expending our bombs on, so we hit our alternate target, San Giovanni, a small town about halfway between the southern front and the Anzio beachhead. We started our dive-bomb run at about 10,000 ft, released our bombs at about 2500 ft, and then zoomed back up to about 7000 ft. The flak was pretty heavy in the bomb run and consisted mostly of 20 to 40 mm, with a smattering of the famous German 88 mm. We had eight bombs that were direct hits in the center of the town, and the other four were in the outer perimeter. This was considered a very successful mission.

As the flight was rejoining after the bombing, my engine suddenly quit. We were about 25 mi inland from the sea, so I immediately headed for the water. Every once in a while, my engine would start and run for about 5 seconds and then quit cold again. I was losing altitude rapidly and was only hoping I could get far enough out from shore before I bailed out. I was down to about 4000 ft over the coast and felt like I had it made. When I got down to about 2500 ft, I figured I better get out or I would be too low to have much of a chance of making it. I was about a mile from land by this time.

The big danger of bailing out of a P-40 was hitting the tail. A few days before, one of our guys had bailed out very successfully and said he had pulled the nose up to slow down almost to a stall; then he rolled the plane over on the left wing and went out the left side. The guys that were with him said he cleared the tail easily.

I figured I would use the same method. I followed the same procedure he did except I let the plane get too slow before I went out the left side. The plane was stalled, it was falling through on its left wing, and I got pinned under the fuselage about 5 ft behind the cockpit. I knew I was going to hit the tail if I ever got loose, but I had no choice. When the plane's nose got below the horizon, the slipstream blew me off the fuselage. My left leg glanced off the vertical stabilizer and my parachute, which was a seat pack, hit the left horizontal stabilizer. This threw me into a violent tumble, which I tried to control. I had no sensation of falling as I could not tell up from down but realized I had to be going down and better pull the rip cord even if I didn't have the tumbling stopped. When I pulled the rip cord, nothing happened for about a second, which is a long time considering the situation, but then I saw a puff of white between my legs and the chute streamed out in front of me. This meant that I was falling in a prone position with my back toward the ground at that particular instant. When the chute billowed out in front of me, I became the cracker on the end of a very large whip. My knees came up and hit me in the face, cut my lips, and broke my goggles. However, I was extremely relieved that it had opened.

I found out later I was fewer than 500 ft above the water when my chute opened. I only made about two swings in my chute before I hit the water. Four of my buddies stayed with me for a while but had to leave soon for lack of fuel. Then a flight of British Spitfires took over the job of keeping me spotted until air-sea rescue could pick me up. I was in my one-man dingy for about an hour when I saw a large ship coming over the horizon. It was the U.S.S. *Kendrick*, an American destroyer, designated DD-612. They put out a small boat to pick me up and as we got back on board the destroyer a British Walrus flying boat flew over and also an American PT boat from the Anzio beachhead area was coming over the horizon to pick me up. We had great air-sea rescue, which made us all feel good.

As soon as I got on board, they took me to their doctor, who checked me over and bandaged the cuts and scrapes I had received. They put me in a small stateroom and told me to rest up for a while. They came back about an hour later to check on me, and the ship's laundry had washed and pressed my uniform and even had my shoes dry. All of this had happened prior to noon as I had been on the early morning mission and it was now about lunchtime on the ship. The menu that day consisted of sauerkraut and wieners, but they didn't think I would like that so they fixed me a meal of baked ham, boiled potatoes, creamed spinach, and blueberry pie for dessert. I didn't know such food existed in the war zone.

I was told that we were on the way to Naples harbor, as the admiral in command of the fleet of destroyers wanted to meet me because I was the first live pilot they had rescued. He was at anchor in Naples on board their flagship, which was the cruiser U.S.S. *Philadelphia*. We steamed into the harbor at about 1400 hours and I was transferred to the flagship where I met the admiral and had a nice long talk with him. While we were talking, he asked if I would like a chocolate sundae or something for a snack. I told him I would bail out again for a chocolate sundae. That was a real treat I'll never forget. He was very nice and invited me back on board anytime I was in Naples.

After our chat, I was getting a little tired and they put me in a cabin to rest. I told them I probably wouldn't be hungry for supper. At about 1830 hours, a mess boy came in with a piece of pumpkin pie and cup of coffee. I got up a little while later and went up to the wardroom and talked to fellows until about 2000 hours, when they had a movie scheduled. After the movie, we all talked for a while and I asked if I could have a cup of coffee before I went to bed. They said that they had saved a plate for me from supper, so I ate my supper at about 2230 hours. It mainly consisted of a large steak, which was a great surprise and very much appreciated.

I asked where I was going to sleep and they took me to the admiral's combat quarters, which were on the bridge. He only used them when they were under way or in a combat zone. They were a lot more comfortable than the GI cot in the pyramidal tent I had been calling home. The next morning for breakfast, I had two fried eggs, bacon, and fried potatoes. The greatest thing about the Navy is that they always have good food no matter where they are or what they are doing. The guys from my squadron showed up on the dock at about 1100 hours to take me back to our base, and that was the end of this experience.

My Longest Day

Told by Roy D. Simmons, Jr.
Captain, U.S.A.A.F.

May 26, 1944, is a day I will always remember. As a matter of fact, it had been just 2 years to the day since I raised my right hand and said, "I do." Becoming a pilot in the Army Air Force and wearing those silver wings was a dream come true, and with those wings came hours of enjoyment, terror, and boredom.

Flying a F-6C/D (P-51C/D) and being assigned to the 111th Tactical Reconnaissance Squadron in Italy, France, and Germany was my combat contribution during World War II. As tac recon pilots, we were instructed not to engage the enemy, if it could be avoided, and not to strafe, but to visually observe and photograph enemy movements, reporting such observations that would enhance the cause of friendly forces. Often while over enemy territory, we would give target coordinates to the Army. They would then place artillery fire on these targets, and we could observe and report the results. But who could pass up a "juicy" ground target? So frequently, we would expend our ammunition on targets of opportunity.

My story took place on May 26, flying from Santa Maria, Italy (20 mi north of Naples), in *Flying Jenny* (my airplane). My wingman and I were briefed to conduct a reconnaissance mission (sortie) near Anzio beachhead and then land at Anzio to be refueled. We were then to proceed to another target area and adjust artillery fire for the Army, again land, refuel, and return to Santa Maria. This would be a very long day.

Arriving over our assigned area, everything seemed to be going in accordance with our intelligence briefing. The U.S. Army was shelling the Germans. The Germans were shelling the U.S. Army. Some ground movement and a small amount of flak were observed, a routine mission. My wingman seemed alert; his assignment was to clear the area and watch my tail while I concentrated on the assigned mission of observing enemy ground force action, taking notes, and giving timely reports to the Army. Then all of a sudden, all hell broke loose. Behind us and rapidly approaching were 12 ME-109s intent on shooting down these two dumb reconnaissance pilots. But this was not to happen. My wingman was still not aware of the ME-109s, so I waited until just before they got into firing range and called out, "Left break *Now!*" Needless to say, I had a startled wingman. Now we were facing 12 ME-109s with airplanes going in all directions.

My immediate reaction was to get the hell out of there. The option of hitting the deck was out as there was a solid layer of bursting flak beneath us, looking like a cloud. Where was my wingman? I saw an ME-109 rapidly closing in on him, so I pulled over and gave the 109 a short burst, observing strikes. Looking around, I saw an ME heading for me, and as we were near the beachhead, I called out to my wingman, "Let's head for Anzio." I then pulled what we called the "emergency tit," causing *Flying Jenny* to really rev up, resulting in the MEs giving up the chase and heading toward Rome. All during this fracas, our ground personnel at Anzio had been listening to our chatter and had begun preparations for our landing. My airplane suffered no damage; however, my wingman had received a hit in the right wing root section. The damage was slight and did not hamper further flights that day.

After our landing and debriefing, we were again briefed for the artillery adjusting assignment, our second mission of the day. This mission was rather routine, actually boring in a way, with me calling out targets by coordinates and then the Army placing artillery fire on the target. After hitting several bridges and having been airborne for a considerable period of time, we again landed at the beachhead, were debriefed, refueled, and planned on heading for Santa Maria.

While being debriefed by intelligence, a request came in from Army headquarters for a recon assistance run over Fondi and Gaeta Point. Our return flight home would normally be out over the water, bypassing Gaeta Point. So yes, we would take this mission as Fondi and Gaeta Point were in the general direction of Santa Maria.

A delay was experienced prior to our takeoff. A seriously damaged Spitfire landed and stopped, still on the runway in front of our parked aircraft. Immediately behind him a damaged P-47 came in landing hot, without brakes, and struck the Spitfire, with the pilot still in the Spitfire. No one could tell if the pilot had survived or not, and contrary to all I had ever been told, I did a no-no and ran to the damaged Spitfire. A number of others and I began removing the pilot and in doing so tripped the gun switches. The Spitfire's guns began firing. My position was close to the left guns, so close in fact that I could feel the heat as the guns were firing. Another few inches closer and I would not be writing about this incident. One of our pilots, who stayed away from the accident, was standing on the wing of an aircraft, across the runway from the Spitfire, when the guns began firing. One of the shots hit his arm. The arm later had to be amputated.

After the runway was cleared, we headed for Fondi and Gaeta Point. Little was going on in the Fondi area, but for Gaeta, well, it seems the "Point" was never very hospitable toward reconnaissance flights. This mission was no exception, as we were met with a blanket of flak. Seeing very little enemy movement, my wingman and I called it a day and headed home.

My longest day was about to end. From dawn to dusk, I had flown three missions, engaged in aerial combat, adjusted artillery fire, and witnessed and became involved in an aircraft accident and nearly gotten killed. As I was leaving operations after being debriefed following the third mission, I checked the mission assignment board. Yes, there I was scheduled again for an early morning takeoff the next day.

A thought crossed my mind as I was going to the bivouac area: 39 missions in less than 2 months. Perhaps I would be heading to the United States in a short time. But this was not to happen.

Time to say good-bye to May 26, 1944, to *Flying Jenny*, to the war, and get a good night's sleep. Thus ended my longest day.

Normandy Revisited

Told by Joe Thompson, Jr.
Major, U.S.A.A.F.

A few years ago, I wrote a paper describing a specific reconnaissance mission on June 4, 1944. The last photograph
was to be of Grandcamp, on the coast of Normandy—the larger area, with a designated spot now known as Pont Dehoc,
overlooking Omaha Beach. At the time of the mission, I had
no inkling of the scheduled landings and justifiably so, for no
one flying over German-held territory prior to D-Day was given
such precious and potentially damaging information. Suffice it
to say that as we and other recon squadrons photographed the
entire French coast, verticals and obliques, the experience was
somewhat like the clay pigeon at a trapshooting meet of top
skeet enthusiasts.

On June 6, the weather was poor. One of our younger
pilots was given the difficult task of flying down the
Cherbourg peninsula to the fishing village of Granville, where
the direct rail line from Paris terminated. We knew the
Germans would bring their panzer divisions by rail and then
move them up to the beachhead. Charles "Stoney" Stone of
Linden, Texas, would be flying VXT, my plane, which had just
made the June 4 photos.

Stoney didn't make it back. In fact, he didn't even get to
check the rail line. The German flak was as deadly and heavy as

any location we later encountered. I had that same mission 6 days later with Remy Chuinard, to whom I had earlier written as part of the Normandy revisited plan.

Remy was 10 years old that murky June 6 morning. He was peering out of his mother's third-floor window overlooking the harbor, witnessing the tremendous barrage of antiaircraft fire. He saw the P-51 flash across the sky, receive the impact of several rounds, and start to fall. When Remy first wrote to me in November 1993, he told of seeing this horrifying end of a young Allied pilot's life. He knew his name and where the plane had crashed, about 2 mi northeast of the town. In fact, he knew the names of all the pilots we lost in Normandy, as well as some of the fighter bombers. His reaction to the liberation of his country became a pursuit of all those who had died and those who had survived as well.

When the D-Day 50 years later celebration began, he was sought out by nearly a score of families from America, some coming to see the area where their loved one had crashed, others accompanied by the pilot who survived. They walked again through the tiny village which had sheltered the flier until the FFI could smuggle him out to the coast and get him safely back to England. He proudly showed me his "library" with volumes of photographs and notations, and letters from all the American families he had received. He was a modest, soft-spoken man, now with an impairing heart ailment. I had anticipated the emotional depth of this dedicated Frenchman, and at the end of our visit, I presented him with a 1994 American silver dollar. He could not speak. He wept.

I had one mission on June 7 and two on June 8. This meant I had a mission on my birthday! The first one, at 0400, was to monitor that same rail line, hoping to catch German troop movements before they could get under camouflage cover. We were flying from a base in England near the coast, but there was the hazard of 140 mi over water before reaching the target area. That's like driving the interstate from Nashville to Cookeville and then 60 mi further. Weather was still terrible and the tacticians wanted to know of any troops near Bayeux (of tapestry fame) and St. Lo, the larger city a few miles further inland. St. Lo was the block in the route out of the beachhead; it was almost destroyed over the next several weeks.

Back from the flight, I was just in recovery when intelligence called again. "But," I said, "I was just over there early this morning." The intelligence officer went right on explaining where additional Nazi troops might be moving up, Vire and Coutance being towns a bit deeper into Normandy. I remember thinking, as my No. 2 and I taxied out, "This sure is a helluva way to celebrate age 25."

Missions continued at a relentless pace, nearly one every day except the 12th, but two on the 18th. One intriguing problem involved a "disappearing German gun" which was devastating our landing troops. Our battleships could accurately spot where the round came from, but when we got there, there was no target! It turned out to be a railway gun moving back and forth and ducking into a short tunnel, only to come back out the other end and fire again. One of my colleagues, Lieutenant Louden, discovered the gun in the act of firing and saw it head for the tunnel. He was flying quite low and the light flak got him. The round exploded in the edge of his cockpit filling his back-type parachute with shrapnel, with a few slugs in his shoulder. One tiny sliver slit his eyebrow, but didn't hit the eye. Louden was a stubborn Dutchman from the wildcat oil fields outside Kittanning, Pennsylvania. He wasn't going to bail out and lose his plane, and with blood streaming down over his left eye, he didn't feel very cheery about flying back over 140 mi of water to our airdrome there. He saw a short landing strip on the beach area for C-47 cargo supplies, not nearly long enough for fighters. Louden ignored the suggestion of his wingman to land wheels up. He circled and, with only one eye, put down his wheels, full flaps, and landed that baby safely. They gave him the Silver Star for that!

After one last mission on June 29 from our base in England, a strip having been built near Bayeux, on July 4, 1944, I landed for the first time on A-9, at LeMolay, a small village near Bayeux. "Lafayette, we are back."

Ours was the first major airstrip in the central beachhead and only 4 mi from St. Lo. We had to be careful in our flight pattern or the Germans would take a potshot as we came in to land. Kept things interesting.

The *Front Row* photographs, which some of you had a chance to observe, were my main outlet, as an escape from all that was happening to me and my fellow pilots. So it was logical to photograph the French people and the Norman apple orchards. I did make a rather first-line photo of Louise Marie and her daughter using our photo lab facilities. Then Patton broke out west of St. Lo, and in a few days, we were gone to Buc Airdrome near Versailles and on to Belgium. But 50 years later, since Caen and Nashville were sister cities, the exchange committee requested a copy of Louise Marie's picture. They would find her and make her the centerpiece of the 50th anniversary of Normandy and the town of Le Molay! I was somewhat disbelieving. "You will find her, after 50 years," I said in October 1993. "Oh, we will run the photo in the weekly paper and the family will call and say that was grandmother. She died 10 years ago. Or she may call, herself." They ran the photograph in the paper. The phone rang and when the editor answered, a voice said, "C'est moi!" Then Louise Marie wrote a three-page letter in French. She had wondered if I had survived the war. She went to the cemetery on Normandy beachhead. No one could give her any information! Her granddaughter enclosed a post card, written in English, which said, "My grandmother waits, with impatience, for you to come back." I said, "Martha, we'll have to go, so let's go early and leave before June 6, 1994!"

In May 1994, we did return to Le Molay, the little village almost entirely rebuilt, and were driven to Denise, the daughter's modest stone cottage. There were over 40 relatives in the farmyard as we arrived. It was an emotional reunion. Louise Marie looked 50 years older. So did I. Soon a typical French dinner ensued along with many tearful embraces and countless photographs. Each serving of a single item, with wine, was followed by another serving, with a different wine. At 2230, we had not reached dessert, and it was really 0300 by jet-lag time. Martha was falling on her face! I arose. "Mes amis de Normandi c'est necessaire dormir bonne nuit!"

The Normans of that little village are truly grateful, although many civilians died in the weeks of the invasion and struggle on the beachhead. Madame Marie and her family survived by leaving their farmhouse with food and clothing and hiding in the hedgerows and ditches until the Allied troops pushed past.

By the end of July 1944, missions continued, and we were flying deeper penetrations to discover backup German troops and tank columns. Then the proverbial "close call" came along. I was "training on the job" a young, not so sharp pilot, Lieutenant Sargent. He should have stayed a sergeant because the sergeants really run most military units. The officers think they do, but the sergeants know it all! At any rate, he was flying No. 1, supposedly navigating and searching for enemy movements. I was watching for flak and enemy fighters. When I saw the same little village pass under my wing a second time, I knew Sargent was lost. "All right, Red 1, I'll take over. Let's go home." Too late—we flew over a six-gun heavy flak battery. The impact of the first close bursts shook my plane, and the engine stopped. I automatically dropped the nose of the plane to lose altitude and avoid further bursts and pulled back the canopy, ready to jump. It is necessary to pull off helmet, radio lines, and seat belt, but at that moment, the engine started up again. I closed the canopy, slid sideways to avoid gunner accuracy, and called in a Mayday. The beachhead was 10 minutes away. Sargent was off to the side and didn't get any hits. Just his luck!

I came in without a landing pattern and slipped in toward the runway. At 50 ft above the ground, the engine failed entirely. I landed dead stick. The flak had cut the hydraulic line to one brake, so we ground looped and parked neatly to the right, 10 ft from one of those famous Normandy hedgerows. My crew chief told me, proudly, there were 130 holes in that plane. There were some other close calls, but that was the closest.

We were paid in French invasion currency, printed and paid for by the United States. This was to provide funds for the inevitable poker games. At night, there was one big tent with a single large light bulb, and a loud-voiced gasoline generator provided electricity. Everything else was by personal flashlight. This particular night, the poker game was in high gear. The game was "baseball." For the uninitiated, baseball rules were that 3s and 9s are wild and a 4 gets an extra card. It was easily possible to have 7 aces in your hand. In the midst of a sizable pot in the center of the table, with 7 players and 10 onlookers, came the unmistakable sound of a German bomber en route to our beachhead sector. The Germans, precise as they were, came over every night at 2300 and 0100. You could set your watch by their arrival time. This was a JU-88, the best twin-engine plane in the war, with hulking, powerful engines always out of sync. "Listen," said a player with a poor hand. "Isn't that a German? We'd better call this game off." "Nah," said the man with the best hand, "that's one of ours." We had a 90-mm antiaircraft unit protecting our airstrip, and at that moment, the guns cut loose a terrible barrage. Someone leaned over the table and turned off the light. The tent swayed and 14 hands grabbed for the pile of bills in the center of the table. We all ran for the foxholes. Exploding shrapnel from ack-ack fire does come back down, you know!

Our squadron flight surgeon, Doc Trimble, had persuaded our tent to dig our foxhole just a bit longer to hold him too. He argued that if anyone was hurt, he'd be right there. He was also lazy! He and I hit the foxhole at the same time. All safe. I could hear Doc in the darkness counting to himself 10, 20, 30, 40, and he wasn't even a player in the game.

After my first letter from Louise Marie, I received a letter in French from a man whose name was Jean Mombrun. He was very excited over our planned visit to Normandy beachhead and Le Molay. He volunteered to meet our plane at Orly Airdrome in Paris and drive us to the Marie home. He wanted as many photographs as possible of our beachhead struggles. He had been 9 years old and was much impressed with the heroic American troops. He was, he stated, organizing a museum in honor of the 50th anniversary. I must have sent forty or fifty 11×14 enlargements and he continued to plead for more. What I did not realize was that while he intended to have no admission charge, he plotted to have me cut the ribbon at the grand opening. He had an associate selling five post cards in a pack (reproductions of *my* photographs, including one of me), and I was held as the main attraction to autograph these "souvenirs." The French are great on souvenirs! The whole village turned out. It was a zoo!

Later, he took us on a tour of the beachhead driving his car faster than 60 mi/h and spouting French at the same rate. Yes, we walked Omaha Beach and saw on the ground what I'd only seen from the air before. It made an emotional impact. Almost more than the impeccable cemetery, and all those crosses and an occasional Star of David, was an unscheduled visit that really pulled at my heartstrings. En route to Omaha Beach, Jean suddenly pulled into a small park just off the roadway and was uncharacteristically silent. I climbed out of the car, telescopic camera in hand, wondering why we'd stopped. Then I saw four heavy German gun emplacements, solid concrete protecting breastworks, long gun barrels still pointing skyward, though the fourth one had taken a direct hit and was destroyed. These were the guns we hunted for, skillfully camouflaged, and which had sent devastating fire, volley after volley, toward the beach and our landing craft so long ago. I stared at them as though they might begin to fire again at any moment!

At that precise moment, two yellow school buses just like ours rolled up, and about 75 fourth-grade boys and girls burst out and began to run up on the concrete emplacements. The boys were showing the girls how quickly they could get to the top of the gun position. Others, especially the girls, were picking the wildflowers that grew in profusion along the path and fence beyond the guns. They were greatly enjoying this brief school outing. I thought, "What a paradox. These guns stand as a stark reminder of the ragged edge upon which Western civilization hung, and these naive children represent the new chance now available, because the guns failed to deter our landing troops."

Then one of the teachers saw me gawking at the youthful energy and happiness, looking every bit the part of an American tourist, camera in hand. "Children, children, come over here. Remember we were talking about freedom last week? You have freedom today because this man and thousands like him came to give it back to you!" Shiny, pink, freckled, tousled faces came up and stared. And they began to try out their English. "Hello." "Good-bye." "Nice day." "How are you?" "Come again." Martha and I have a treasured photograph as we stand before one of the disabled guns surrounded by 60 French schoolchildren, who were having a wonderful time together.

We flew the missions that set up the breakthrough west of St. Lo. I witnessed the heavy bombers dropping a devastating carpet of bombs that launched General Patton's swift penetration, and in a few days, the tanks were beyond Chartres, and Paris was liberated. But that is another story.

Back at Le Molay, 1994, the mayor of the town, who spoke no English, pinned a medal on my tunic and I was required to make a response to 300 townspeople who had gathered in the only meeting hall for the occasion. "Mes amis du' Normandie, c'est une grande plaisir parle avec vous c apres midi." I made a big error in my effort to speak their language. Remember, Louise Marie was 23 when we first met at the A-9 airstrip in July 1944. I pointed to her and (correctly) said, "Dans la guerre elle est vingt trois ans." (She was 23.) Then I said, "Maintenant [pointing] elle est quartre vingt." (Now she is 80.) This brought down the house! Never make a lady older than she is!

We gave Louise a sterling silver seagull on a silver chain, symbolizing that freedom and peace had come from the sea! Jean Mombrun drove us to the airport. We had been there only 6 days, but it was most memorable! He could not say what was in his heart, he spoke no English, but he thanked us profusely in French. There were tears in his eyes.

D-Day

Told by Enoch B. Stevenson, Jr.
Major, U.S.A.A.F.

Since I was the operations officer, I alternated with the squadron commander. I'd fly four or five times a week, then maybe once or twice a week. The rest of the guys would fly five times a week every week. The only thing that would alter our routine was the weather in England, which was so bad it kept us from flying much of the time.

We didn't have nearly the rules they have now; there were none of the crew rest requirements or anything. Preflight in those days consisted of kicking the tires. The P-51 had a liquid-cooled engine and was vulnerable to groundfire. If the Germans got one in your radiator, you couldn't last; you couldn't fly anymore. If that happened, all you could do was turn toward friendly soil, fly the plane until it quit, and then bail out.

We knew D-Day was coming, but we didn't know when. Then, on June 5, 1944, we flew and got back in the late afternoon. As was customary, since we didn't think we had to fly that night, we decided to go to the bar and have a little attitude adjustment. When we got there, they told us the club was closed and everyone was restricted to base. They told us to go to our quarters and get some sleep. As we went back to our quarters, we saw the group commander walking around with his superior, a brigadier general, both with maps and charts under their arms. It didn't take a rocket scientist to realize it was D-Day.

We didn't get any sleep, and at 2230 hours, they came and told us we would brief in 30 minutes. So we went flying, having already flown that day, and having not gotten any sleep. The squadron, about 20 planes, took off at 0030. I was leading. We were nervous at takeoff. But the most nerve-racking part of that D-Day flight was the night takeoff, which we weren't accustomed to doing. We had 20 airplanes and had to take off and then get into formation. The weather was terrible and the ceiling was down to around 500 to 800 ft.

Under normal daylight conditions, I'd have made about two circuits of the field, and by that time, everyone would be in formation and we could head out on course. But under those circumstances—dead dark, low ceiling—all you had to go on was everyone else's lights.

That morning, I made one circuit of the field and I heard guys saying, "I can't find you! I can't find you!" and I said, "All right! Let's calm down, now! Just head out on course." We headed out to the south a ways. And it wasn't very long until everyone was in formation.

We didn't have any bombs because we were using external fuel tanks instead. That way, we could stay in the air longer. We had six .50-caliber guns that had incendiaries and tracers mixed in with regular bullets. Our mission was to ward off any German interference with the operation. First, we flew a patrol route, back and forth, between Land's End at the end of southeast Britain and the Jersey and Guernsey Islands off of France. We were underneath the overcast, which over the ocean was about 1500 ft.

It was as black as the ace of spades when we took off, but daylight started early because of daylight-saving time and because we were so far in the northern latitudes. We saw nothing of the invasion, no ships or any of that. We were well enough away from it and were down too low to see it. We expected to see some German fighters coming out to attack the invasion, but the Germans didn't put up any air resistance at all. Our participation went very smoothly.

We flew patrol for about 3 hours, and then at 0330 we were relieved by the other two squadrons. We headed east behind the invasion beaches and looked for targets of opportunity. We found a German convoy of about a dozen trucks down around Caen, France, headed toward the beachhead. We knew they were Germans because, for one thing, it was France, and for another, we could see swastikas on the hoods of the trucks. We pealed off and got into trail, one behind the other. Since I was the leader, I was in front. When the Germans heard us coming, they stopped in the middle of the road, got out of the trucks, and ran for the ditches. We destroyed the convoy and headed back to our base in England. We landed at 0530 or 0600 hours. We refueled, rearmed, and had some breakfast. Then everyone else took off and did the same mission again. I didn't fly on that second mission; the squadron commander did.

The D-Day missions were a little unusual. Normally, we flew escort for heavy bombers, B-24s (liberators), and B-17s (flying fortresses) on their daylight air raids. We were faster than the bombers, so we'd fly on top of them doing an S-pattern. They flew at about 25,000 ft; we at about 26,500. Since we were so high and since aircraft weren't pressurized, we'd be on oxygen the whole time.

The Germans considered the bombers, not the fighters, the enemy. It was the bombers that were destroying their cities and their infrastructure. So the Germans weren't aiming at us. Of course, some of our guys got hit by flak, but the target of the flak, and of the German fighters, was the bombers.

As the Army started working its way across Europe, we continued to escort the bombers. But after they had completed their bomb run and were headed back to England, we'd leave them and go look for targets of opportunity. The primary target of opportunity was an airfield loaded down with German fighters. They didn't have any fuel by then and were always on the ground. We had effectively destroyed their capability to produce airplane gasoline by then. In fact, we shot a lot of airplanes that didn't even burn because they had no fuel in them. Every once in a while, they'd find some gas and mount an attack, but basically, they were out of business.

Another primary target was a locomotive, and boy, they were great targets. You'd hit one and they'd go phew! There wasn't much you could do with roads or bridges, unless you had some 500-lb bombs. On occasion, we'd carry two bombs, but carrying those considerably restricted our range. Our bomb sight was a fixed gun sight. You aimed, released the bombs, pulled up, and that was that. You only got one shot.

Where Were You on D-Day?

Told by Leslie E. Traughber
Captain, U.S.A.A.F.

When asked this question on D-Day, I only had a vague idea as to where I was and what I might have been doing on June 6, 1944. I knew I was in Italy and was a flight leader participating in one to three combat missions daily. Rome had fallen and we were hitting the German tanks and trucks retreating north. We were having a great time bombing and strafing everything that moved. Most of our flights were short and sweet. Hit them hard, return to base, give a report, load up, and return for targets of opportunity. But as enthusiastic as we were in pursuing the retreating forces, our losses were beginning to climb. One or two planes out of every four were either lost or damaged on each mission.

I was aware the invasion of France had begun and most of us did not consider it anything different from what we had been doing for the last several months. The 27th Fighter Bomber Group had already been a part of the North Africa, Sicily, Salerno, and Anzio landings, and we had just broken out of the Anzio beachhead. During the month of May and throughout most of June, our fighter squadrons of the 27th Fighter Bomber Group were flying P-40s which had recently been brought out of mothballs while we awaited the arrival of new P-47s. We considered the invasion of Europe to be one of the reasons we were all there, and those of us who were active in daily combat were trying to stay alive. We were much more interested in happenings that directly affected us.

Since then, I hadn't given the Normandy invasion much thought. I researched my file and later called my friend, Charles Waddell, who had been with me during this time, and it was he who reminded me of a story that I have always found fascinating and intriguing.

I had never thought I was superstitious but I believe most fighter pilots are very methodical about their flight preparations. There were certain habits we formed, such as putting on the same glove or the same shoe first, and these habits were repeated daily until they became so entrenched in our subconscious that we would not change our routine for anything. The only time I can remember changing my routine was the day I got shot down.

Approximately 8 or 9 days before D-Day, we had finished the day's missions and the four of us—Charles Waddell, John Wagner, Bob Schulte, and I—were relaxing in our tent. We had been together through advanced flying school at Eagle Pass, Texas, and combat flight training at Hillsboro Field in Tampa, Florida, before going overseas, and we had been living together in the same tent for several years. We had experienced a long and close relationship. On this particular night, our special service officer dropped by. He had never visited us before nor did he ever visit us again. We were having a regular bull session when we learned one of his hobbies was reading palms, so we asked him to give each of us a reading.

I was the first to have my palm read and was told I would have a life-threatening event, but I would survive it without consequence. This happening I interpreted to be the time when I had been shot down earlier in my tour of duty. I had parachuted to safety and returned to duty on the same day. Charles Waddell was told he would complete his tour of duty without any difficulty, which he did. The officer implied John Wagner would have a problem but he would also return home safely. Bob Schulte was told something that I cannot recall, but it was of no consequence at the time. The officer read many other things in each of our palms, none of which were given a second thought. Wagner and Schulte later left the tent, and as the officer, whose name I cannot recall, was leaving for his quarters, he told Charles and me he was unable to tell Schulte what he saw in his palm because he only saw death.

A week or so later, on June 4, John Wagner was flying on a strafing mission north of Rome. He was reported to have been last seen at 100 ft above the ground and going in. We later learned from John that he bellied in and was picked up by the 88th Infantry Division. He remained with them for 3 or 4 days and observed them overrunning a German roadblock. He was soon assisted in returning to Rome. On June 5, Bob Schulte was reported to have been unable to pull out of a low-level strafing run on a motorcade north of Rome. His plane flew into the side of a mountain and exploded in flames.

Charles and I were the only two remaining in our tent on June 6, D-Day, after the clairvoyant experience with this special service officer who now had become questionably special in our lives. As bad as we felt, we also felt reassured that we might make it home.

About 10 days later, we received a message to meet Lieutenant Wagner in Rome on the steps of one of the monuments. All was well with John, as he had been enjoying his freedom and the amenities of a liberated city for a few days before reporting his whereabouts. He was happy to be back, but he never forgave Charles for giving all his belongings away, especially his cigars.

I don't believe it was anything but a coincidence that these memorable events took place after their predictions. But after returning home from overseas, a friend and I were returning home from Valdosta, Georgia, after having been on a weekend pass from our post in Perry, Florida, when we passed a palmist in a trailer home by the side of the road. I had told him about the special services officer, so he wanted to stop and get his palm read, which we did.

While there, I had mine read also. This time, I was told I would be married, have children, and be very happy, but that I should be aware of a tall dark stranger who would disrupt my marriage. As you might expect, I am still looking for that "sucker."

Zero

Told by Frank N. Davis, Jr.
Lieutenant Junior Grade, U.S.N.

Since World War II, we have heard of the maneuverability and vulnerability of the Japanese Mitsubishi Zero (Zeke) fighter and that no American fighter could outmaneuver the Zero. In relating an experience of mine, I do not intend to imply that I was capable of more than other pilots, but rather what a Wildcat FM-2 was capable of. In 1943, the Wildcat was one of the older fighters designed by Grumman and designated F4F. Later, General Motors assumed production of the Wildcat. It was updated with a new engine and other modifications and designated FM-2, a hang tough fighter, which was primarily used on small aircraft carriers called baby flat tops or Jeep carriers.

On June 19, 1944, when the "Marianas Turkey Shoot" was winding down, two divisions of Wildcat FM-2s, eight airplanes, were launched for the early morning combat air patrol over our carrier task force. Bogies (unidentified aircraft) had been reported in the area south and east of Tinian. Two of us had been vectored (directed out on two or three headings), and our altitude was about 18,000 ft when we were directed out again on bogies at two o'clock high. We initiated a climbing interception. As we climbed, the bogies were identified as two Zeros at about 20,000 ft. We dropped the wing tanks and added power in preparation for the interception and engagement. My engine ran rough and missed at the higher power setting, so I had to reduce power to about 85 percent to smooth out the engine. I was dropping behind and below my cohort, Perry. The Zeros first turned toward us and then headed away as if they were waiting for us to get in position so they could dive down for an attack.

As we were climbing and closing, I had a short time to consider my situation. I was flying an airplane that was not supposed to outmaneuver a Zero, even under ideal conditions. My engine was not performing as needed and desired for a combat situation. At 20,000 ft, I was above the most effective fighting altitude of this Wildcat, as it begins to get mushy and not as responsive in maneuvers. The enemy aircraft had the altitude advantage and could initiate the attack when desired. I was left behind for a one-on-one situation and could not provide or receive the mutual support we both needed. However, regardless of the circumstances, I would still prefer to be in the Wildcat. The odds did not appear to be in my favor, so I figured it was about time for a short prayer.

Perry got close enough to fire at the closest (second) Zero, and then they fell off into kind of a loose wingover with Perry in pursuit. The lead Zero came in to help No. 2 by firing at Perry, and by this time, I was about to arrive on the scene. In order to distract No. 1 Zero, I fired a burst that fell short, as I was too far away. He did not like the tracers going under his wing and elected to do a split-S with me right behind. The Zero continued in a tight circle with me trying to get in a position to fire. I fired again but fell short because of the tight circle and the G forces with the tracers making a rainbow arc falling from the wing guns. Continuing the tight circle, I knew I had to get a considerable lead with a tighter turn inside the Zero to compensate for the G forces and the turn, so I pulled back as hard as I dared with the Wildcat shuddering and the canopy rattling on the verge of a high-speed stall. We continued the tight circle, and on the next firing burst, the tracers appeared to be going into the engine and fuselage.

The Zero began to flatten out. I fell in behind with all guns firing. The Zero began smoking and what appeared to be a piece of cowling headed directly toward me, so I pulled up sharply to the left to avoid hitting the object, and when I dropped my right wing to look down, no Zero was visible, only pieces. My Wildcat had accomplished in about 60 to 90 seconds what had appeared to be an impossible mission. I then looked for Perry to see if he was okay, which he was. In his report, he saw debris falling, and the No. 2 Zero stayed intact in going straight into the water. Later, upon landing, folding the wings, parking in on the bow of the flight deck, and cutting the engine, the bull horn blared, "Ensign Davis report to the captain on the bridge." After relating our encounter with the Zeros, neither the ship's captain nor the squadron commander responded but just looked at each other, shook their heads, and walked off.

I have thought through this event many times and recall that old expression, "a blessing in disguise." It just could be that with my engine not turning up maximum power it slowed the Wildcat down enough to reduce the turning radius, permitting me to maneuver more effectively. Perhaps this is the way the Lord planned it.

My First Combat Mission

Told by Hensley Williams
Major, U.S. Marine Corps

I went back to the Pacific war in the summer of 1944 as a fighter pilot and was based on several of the mid-Pacific atolls (Marshall Islands group). We strafed, rocketed, and dive-bombed the neighboring Japanese-held islands, which had been bypassed but had not yet surrendered. Some were still potential threats. Some had airfields and others did not. Some sent up considerable antiaircraft fire and others did not. One never knew in advance, and so there was a reasonable apprehension before any mission. The mission at hand was a low (500-ft) photographic mission of the airfield on the island of Ponape, some 475 km from our base on Eniwetok, a rather long flight requiring an extra belly gas tank. Surprise was to be employed if at all possible. A B-25 bomber was to provide navigation and rescue assistance for one other photographic F4U fighter plane and me.

This was a volunteer mission, which was a bit unusual. Needless to say, there were plenty of volunteers. All of us were young and looking for action. I was the latest arrival from the States, the oldest pilot present, and so a bit on the spot. My courage was at stake in a way. The mission was held up for 2 days by bad weather, so I had a little extra time to think about it.

The mission went normally with no significant antiaircraft fire. I reported to the group commander that the only trouble I experienced was a somewhat sore rear end from sitting on a hard-packed parachute and rubber raft.

P.S. Ironically, my last combat mission (No. 52) was over Japan almost a year later and turned out pretty much the same way.

The Mission

Told by Charles R. Mott
Captain, U.S.A.A.F.

The following words appear in the story you are about to read. These brief explanations are given to acquaint you with the meanings in advance:

Angel: altitude

Angels 35: altitude of 35,000 ft

B-29: largest bomber used in World War II

Bogie: enemy aircraft

Buster: full throttle

Cleveland Green: code name of flight

Cleveland Green Leader: the lieutenant

Dinah: Japanese photo reconnaissance plane

Fighter Control: headquarters for squadron communications

Flight: formation of four fighters

P-47 (Thunderbolt): plane manufactured by Republic Aviation Inc.; largest single-seat plane built at this time

Project First: project for the first bombing of the mainland of Japan from China

An airfield in northern China, July 4, 1944, 0400. Four members of the 58th Fighter Squadron are on morning alert. I was a young lieutenant on my first combat alert as a flight leader. At this time, the pilots are relaxed and half asleep. Nothing happens at this hour. Then the impossible. The alert phone rings, direct from fighter control. I answer the phone, "Cleveland Green Leader." A crisp voice from the other end responds, "Cleveland Green Leader, Angels 35, area 6-A, Buster." I know this means action. I yell at the other three pilots, "Let's go! Take off south to north, Angels 35, Buster."

I run to my plane, heart pounding as I prepare to lead my first combat mission. I check with my crew chief to make sure my plane is full of fuel and ammunition and is ready for takeoff. The crew chief jumps from the wing and yells, "Good luck, lieutenant." I check with the rest of the flight and then give the order to take off.

My tension and concern are increased by the fact that these men have never been to 35,000 ft as a combat team with orders to get there at full throttle. I am certain that with these instructions the enemy has already been positively identified and is fast approaching the area. The flight is airborne without incident. I order, "Close up, let's stay in tight. We've got a long way to go." Every member of the flight strains his eyes and continuously turns and looks in every direction for a sign of the enemy. The second lieutenant, flying the flight leader's wing, hits the transmit button. "Cleveland Green Leader, this is No. 2. Where are we headed? What is the mission?" I respond, "Not sure, No. 2. Will check with fighter control when we reach Angels 35."

The fog is lifting from the rice paddies of China as it has for centuries. The only noise to disturb the tranquillity around this ancient setting is the drone of four P-47 Thunderbolts, the largest single-seat fighter that has ever been produced by the genius of American industry. The tension of the pilots increases with every movement of the altimeter toward 35,000 ft. Finally, the magic number is reached. The flight levels off. I press the transmit button, "Hello, Fighter Control. This is Cleveland Green Leader, am at Angels 35, area 6-A. What is our target?" The answer rings loudly in my headset, "Cleveland Green Leader, this is Fighter Control. Twin engine bogie in area 6-A should be on left and below." Fighter control has both the enemy target plane and the American Thunderbolts clearly on radar and gives these instructions with the authority of the blips on the radar screen. I look to my left and below and excitedly report, "Roger, I have him in sight. Bogie is a twin-engine Dinah." I give instruction to the flight. "Cleveland Green two, three, and four, entrail me. We'll attack in single file. Tallyho!" The Thunderbolts roll out at an interval to ensure that no other member of the flight moves into the line of fire. They move to the attack. "Cleveland Green two, three, and four, this is Cleveland Green Leader. Use a 10-second burst and pull to opposite sides."

The Japanese Dinah, at a lower altitude, does not see the American fighters and fails to take evasive action. The flight leader leads the rest of the flight down to the Japanese Dinah and brings the Dinah into his sights. He closes the range to 600 yd and fires a 10-second burst from the eight wing-mounted .50-caliber machine guns. The lieutenant watches as his tracers hit the crippled Japanese aircraft. Smoke begins to trail from the Dinah. By the time the fourth Thunderbolt has closed in on the enemy, the Dinah is enveloped in flames and is plunging to the rice paddies below.

The flight re-forms at 35,000 ft. "This is Cleveland Green Leader. Check your fuel and ammo and report." "This is No. 2, Cleveland Green Leader. Have 45 gal fuel and 1800 rounds." "This is No. 3, Cleveland Green Leader. Have 40 gal fuel and 2000 rounds ammo." "This is No. 4, Cleveland Green Leader. Have 50 gal fuel and 1600 rounds ammo." "Roger, Flight. Will check back later."

I finally return to the discipline of my combat training and report, "Fighter Control, this is Cleveland Green Leader. Enemy bogie destroyed. What are your instructions?" "Cleveland Green Leader, this is Fighter Control. Remain in area 6-A and search for further enemy contact. How is your fuel and ammo?" "Roger, Fighter Control. Will do. Fuel and ammo okay." "Roger and out, Cleveland Green Leader. Will check back later."

After 10 minutes of further patrol, the flight has not seen any sign of the enemy. The flight is still at 35,000 ft and is unaware of the fog lifting below. "Cleveland Green Leader, this is Fighter Control. Return to base." "Roger, Fighter Control. Will do." "Cleveland Green two, three, and four, return to base." The flight begins its descent. At 6000 ft, the flight is engulfed in the rising fog. I am suddenly aware of the danger of being lost. I hit the transmit button. "Fighter Control, this is Cleveland Green Leader. I am at 6000 ft and lost. Pick up and give vector to base." "Roger, Cleveland Green Leader. You will have to climb." "Roger, Fighter Control. Two, three, and four, I am climbing to establish position and vector to base. Remain at this altitude and course." I begin to climb. I am afraid. This is my first combat mission as the flight leader and I am lost. I fear court-martial for myself and the members of my flight. "Cleveland Green Leader, this is flight control. Steer left 020."

I rejoin the flight and we head for home. Ten minutes later, the flight falls in for its landing pattern. I glance to my right and see the general and the squadron commander waiting for my flight to land. As I bring the Thunderbolt around for the final approach, I am thinking of the humiliation and possible court-martial that could follow my landing. My first combat mission as flight leader and I had been lost!

I managed to make a smooth landing and taxi to a halt. I removed my goggles and parachute and slid the canopy back. The crew chief ran up, jumped on the wing, and jerked me from my plane with cries of "Great flight, Lieutenant." I looked at the sergeant with a frown, still thinking of a court-martial. I jumped from the wing and moved toward the general. My tension was very high because I fully expected a severe reprimand from my superior officers. The general walked up with an out-stretched hand, "Congratulations, Lieutenant. It was a great flight." A look of bewilderment crossed my face. The squadron commander came up to me, "This is the first victory of our fighter wing's mission to protect the B-29s of Project First."

I was suddenly aware that my mission had been accomplished. I had done a good job. I walked across the field. The time was 0700. I got a cup of coffee and headed for my bunk for a well-deserved rest. It had been a long mission.

Rabaul

Told by Beverly W. Landstreet
Captain, U.S. Marine Corps

My squadron was the first all-weather squadron as far as the Navy and Marine Corps were concerned. We wrote the syllabus on the use of radar for navigational and weather purposes. The Bendix and Minneapolis-Honeywell radar were designed to be used for shipping and air-to-air identification of enemy targets. As we determined from experience, the radar was effective for over-water navigation, weather avoidance, and shipping.

Our squadron, flying F4U Corsairs, operated throughout the Solomon Islands from Guadalcanal to Munda, New Georgia, Sterling, Bougainville, Green Island, Emireau to Tacloban, a strip on the Philippine Island of Leyte. We participated in flying night cover for all these landings except Guadalcanal and Leyte. Our squadron's objective was to intercept Japanese aircraft and shipping throughout the Solomon Islands, with particular attention to Rabaul on New Britain and Kavonga on New Ireland. Rabaul was the staging point for the Japanese Navy and Air Force throughout the South Pacific.

One night, I was on station over Rabaul and was picked up by Japanese radar controlled searchlights. I overheard the pilot of a Black Cat (PBY-5A) communicating with a PT boat patrolling the area about me being lit up like a Christmas tree.

He told the PT boat skipper that if I would accommodate them, they would knock out a few searchlights and gun emplacements by lobbing shells on an airstrip that had just turned its runway lights on and off. It was obvious the Japanese were attempting to bring in replacements from Truk or New Guinea, otherwise the strip would be blacked out. The distance from Truk to Rabaul was about 700 kn and about 500 kn from the Japanese-held positions on New Guinea. I began talking to the pilot of the Black Cat since we generally disregarded radio silence, and I advised him that I would fly in and out of the searchlights so they could take potshots at ground targets of their choosing.

The Black Cat pilot said his original mission was to mine the entrance of Simpson Harbor. Since the Japanese had the runway lit up, he might as well lay a few depth charges on it. While they would not detonate, they would be difficult to avoid when and if the Japanese brought in any aircraft. Suspecting that Japanese aircraft were in the area, all we could do was wait as long as fuel permitted and get in their landing pattern and strafe as they landed.

While I was trying to avoid the antiaircraft fire and remain in the searchlights, the PT boat and the Black Cat did their job on one strip. Immediately, the Japanese turned off the runway lights on another nearby strip. The antiaircraft and searchlights stopped. Within minutes, three Betty twin-engine bombers entered the traffic pattern. As they landed, we destroyed them one by one on the ground. I am sure they were low on fuel, as I was, and had no choice but to attempt to make a landing. I returned to Green Island to refuel and subsequently flew to my home field on Sterling Island just south of Bougainville.

I was mustered out of the Marine Corps in December 1945 as a captain and I remained in the reserve until 1958. I flew the following planes, all radar equipped: F4U, F7F, and PBJ. It was many years later that I learned that the pilot of the PBY that night was Joe Cummings, a longtime friend of mine from Nashville.

Turkey Shoot, July 1944

Told by William S. Miller
Captain, U.S.A.A.F.

Bill had an eight-ship detachment of pilots, crew chiefs, armorers, and P-47s from the 91st Fighter Squadron at Ankang, China. They expected to be there about a month in order to work over the railroads and the "few and far between" airfields in that part of China. A similar detachment from the 92d Squadron was also there. They each had two or three pilots from their sister squadron, the 93d from India. Those poor guys from the 93d never did get into the real war except as fillers in the two China-based units. The whole detachment was headed by Lieutenant Colonel Fred G. Hook, who was the deputy group commander.

The squadrons were equipped with the huge Republic P-47 Thunderbolt fighter. The birds were the "razorback" model. This model had a greenhouse type of glass and aluminum canopy, which faired in smoothly with the fuselage, instead of the bulging bubble canopies of other models. In the razorback model, the pilot had to crane his neck a lot more to see to the side and especially to the back where the enemy fighters normally came from. Other than the canopies, the two versions were identical, except that the slab sided version, being aerodynamically cleaner, was about 3 mi/h faster.

The huge 41-ft wing of the P-47 held eight big .50-caliber machine guns with 425 rounds of ammunition for each gun. These guns were upgraded and air-cooled models of World War I guns that could now fire 900 rounds per minute. These guns had a muzzle velocity of 3200 ft/s with great range and penetrating power. Bill always compared the effect of a 1-second burst of 120 fifty-caliber guns to that of driving a concrete truck into a target at 400 mi/h. The big plane could also carry up to a 1000 lb bomb or up to a 110-gal fuel tank in the belly and either a 500-lb bomb or a 75-gal fuel tank on the wing pylons.

The 91st was assigned to fly the first flight that day. It would be a six-ship armed reconnaissance on the railroad leading north out of Hankow. The weapons load was to be a full load of .50 caliber, 3400 rounds. The previous night, Lieutenant Colonel Hook had asked Bill if he could lead the flight.

Fred had only been in China for a couple of months, and this would be his first combat mission. Fred was a thirtyish— no one knew how far into his thirties—handsome, black-haired man who wore a dapper mustache. In fact, he looked and acted like a younger, leaner Clark Gable. He was a former barnstormer, cotton duster, commercial pilot who had paid for his several thousand flight hours by doing anything that could be done, including being a Memphis motorcycle patrolman. He had finagled a job with the Army Air Corps as a service pilot. Service pilots were supposed to fly only trainers and transports in a noncombat role. We called them male WACs. Somehow, Fred had upgraded his pilot status to that of a line pilot and was now a regular fighter pilot. He was the only service pilot any of us had ever known who was able to pull off such a transition.

Bill had first run into Fred a month earlier over Fungwangshan (A-2) at 1000 ft over the airfield when Fred flew down from headquarters to make his first visit to the 91st. Bill and his wingman were returning from a training mission when they heard the 92d call sign. Accordingly, they got into position to jump the outlander just before he was to break for a landing. Bill had called, "92d bird on the break at

Fungwangshan, check your tail at 6:00." Fred had given his P-47 full throttle and one helluva dogfight had ensued at 1000 ft or less over A-2. Considering the fact that Fred was outnumbered and that they had the jump on him before the fight started, he did an excellent job. They had fought for 10 to 15 minutes with Bill's wingman more or less hanging on before they decided that it was time to land. Bill had thought at that time, "This guy may be the third best pilot in the group." Ed Weed was No. 1, Bill was No. 2 by a hair, and this 92d stranger was third. Bill thought, "Give this sucker 20 more hours in the bird where he can make his moves instinctively, and he will really be hard to handle."

An hour after the flight, Bill was summoned to headquarters to talk to Lieutenant Colonel Hook. "Oh, oh," Bill thought. "So that's who it was." Fred didn't do any chewing and apparently only wanted to meet the pilot who had fought him. When they learned they were fellow Tennesseans, they took an instant liking to each other, and in spite of the difference in rank, became good friends for life.

So, in considering whether he should let Fred lead the mission, Bill knew he had the flying skills needed, but he also realized that no one ever knew what was going on during their first four or five combat missions. The next morning at breakfast, Bill told Fred that he had thought about the problem carefully and had decided that Fred should not lead a mission until he had at least four or five missions under his belt. He did tell him he could lead the second element and that he would be designated the alternate leader if Bill had to abort or got shot down. The other people in the flight were Ritchie flying No. 2, George Frieze flying No. 4, on Fred's wing, with Roy Holton and one of the 93d guys making up the third element.

They took off uneventfully and got into a spread combat formation. They headed for the main rail line out of Hankow to Chenchow and Taiyuan. The rail line was about 280 mi, or about 170 minutes of flying time at 230 mi/h indicated. Bill had just briefed them that they would hit the railroad and head northward for an hour or so, depending on how fruitful the mission was. When they hit the Yellow River (Wang Ho), they would follow the parallel railroad line westward to the big bend. From there, they would return to Ankang on a direct heading of 195°. The whole mission would last about 3 hours and 35 minutes. They should return with plenty of fuel if they didn't meet any Japanese fighters. If they met any fighters, they would fight at full throttle, which ran fuel through the huge engine at a rate of 350 gal/h. It would be a new ball game at that point. The P-47s held 370 gal of fuel internally and 75 gal in a belly tank. This gave the aircraft a normal range of 5 hours and 15 minutes, or about 1200 mi.

All pilots had been briefed that if anyone had to abort or return to base, a two-ship element would go home. Alternative air bases had been pointed out in case of bad weather at Ankang. The intelligence officer had briefed them on where the supposedly friendly Chinese nationalist and communist pockets of resistance were in case anyone had to bail out or make a crash landing. Today, their flight color was red and their call sign was eagle. Squadron call signs were normally named for predatory birds or animals or occasionally for weather phenomena like thunder or lightning.

Bill and his armorer had adjusted his guns to converge at 300 yd. He preferred concentrated fire rather than the standard scatter-gun approach. He also made one additional change to standard loading. The normal load was one tracer bullet for every five normal rounds of .50 caliber. He felt that this really didn't work, so he had his guns armed for a steady stream of 10 tracers when he had expended 390 of his 425 rounds. In this way, he could save the rest to fight his way home. It was nearing the fall of the year in China, so the pilots were dressed in their flight suits and leather A-2 jackets. Bill always carried a little 6.35-mm or .25-caliber pistol, which he had found in Italy, in his pocket. This was in addition to his .45-caliber automatic pistol. Sometimes, if they were strafing troops, they also carried a .30-caliber carbine, the better to fight their way out if necessary. Today, they left them home since they did not expect to meet any troops.

En route to the railroad, they passed directly over La Hokow. Bill told his flight to start their necks swiveling so they wouldn't be surprised by any Zeros. This constant neck swiveling against heavy cotton flight suits and heavy leather jackets resulted in very chaffed necks, so most of the boys wore silk scarves for protection. They were also very decorative. They hit the railroad and swung to the north. Bill and Ritchie went down to 1000 ft. Fred and his wingman stayed to the east at 2500 ft. Roy's element flew up-sun at 5000 ft to protect against Japanese fighters. After about 5 or 6 minutes, Bill and Ritchie found the first train. They shot up the locomotive. It was pulling a string of boxcars and they didn't normally waste ammunition unless the pickings were slim. Minutes later they found No. 2. It was a passenger train. Passenger trains were very rare; in fact, Bill found only two during his entire tour.

They made their attack on the passenger train's engine from the rear at about a 15° angle to the track. Their point of aim was the cab right above the engineer's head. This area also held the firebox and the steam boiler. A concentrated burst of .50 caliber would sometimes create an explosion. The burst would also get the engineer, fireman, and any stray "shotgun" riders. Bill's burst was right on, but the engine did not explode. It merely came to a stop. As the train coasted down, Ritchie hosed the cab down with another well-aimed burst. Then Bill instructed Red 3 and 4 to come down and help strafe the passenger cars. The Chinese in occupied areas had been instructed to stay off trains, trucks, buses, and all types of powered conveyance. They had been warned that these were all military targets and would be destroyed on sight by the Allied forces. Any civilians on powered vehicles would be assumed to be collaborators. Therefore, passenger trains were prime targets.

Fred hit the cars next and, as George came in behind him, he called, "Hey, they are all troops in their dress uniforms." It was a relief for all of red flight to know that the people on the train were not Chinese civilians. Bill started his second pass and confirmed that the passengers were all in uniform and none of them appeared to have weapons except for a few swords. He also noted and commented to the flight that the train was on the edge of a 70-ft embankment on the west. That meant the troops could only evacuate to the east. On the east side, there was a 300-to-350-yd level glen. The glen was ringed by semicircular sheer cliffs at least 300-ft tall. "We have them trapped," he said. "They can only get out of the trap by going up or down the track, and we can plug up those holes." Bill called on Red 5 and 6 to come on down and join the action.

Normally, they would have strafed the train from different directions, from various dive angles, and would have been flying erratic patterns to avoid the flak or other groundfire. However, in this case, there didn't seem to be any guns around. Bill asked the other pilots if they had noticed any groundfire. None of them reported seeing any return fire. Therefore, they were told to make passes north and south, parallel to the tracks. When Bill came in for his third pass, he noticed a stream of tracer bullets walking along the ground toward him. He shouted, "Red aircraft on a firing pass from the north, break off." The other pilot replied, "You break off. I was here first." Bill calmly replied, "This is Red Leader and you either break out of the pattern immediately or I am going to shoot you out of it." As he pulled up from his run, he thought, "This is getting dangerous."

In the interest of safety, he told the flight to get into a conventional peacetime ground gunnery pattern and that they should all attack from south to north. They were instructed to call, "Red___ going in hot," as they started their firing run, and to call, "Red___ off," as they turned to the left out of the target area. Bill computed that they had 20 minutes or so before the Japanese Zeros, if any, could get there. However, he warned his flight to clear the skies carefully after each firing run while they were getting into position for the next one.

He thought about letting one of the two ship elements go ahead and fire out all their bullets and return to Ankang for reinforcements." This would be a perfect place for some napalm or parafrags," he thought. Napalm was a jelling agent that could be added to the gas in the belly tank, and it burned with intense heat. Eight tanks of napalm dropped into this glen would cover the whole thing and would kill anything that crept or crawled within it. Parafrags consisted of three 80-lb bombs strapped together. Their cases were made up of 4- to 5-in diameter bands of steel that were welded together loosely. When the bomb dropped, a small parachute opened and the bomb floated gently to the ground. At about 18 in off the ground, a small probe touched the ground and the whole thing blew up. The pilots affectionately called these bombs "grasscutters." They also would be perfect munitions for the present situation.

Bill made some fast calculations about sending back for reinforcements. If his returning element flew at maximum speed of about 360 mi/h, it would take them 35 minutes to get within radio range of Ankang tower. It would take a minimum of 10 minutes to round up and to brief another flight of pilots and another 15 minutes for maintenance to get four birds serviced. If they were loaded with napalm or parafrag, it would take another 15 to 20 minutes even if the napalm or parafrags were ready. Napalm had to be made to order the night before, and it was doubtful if the parafrags would be available at Ankang. He estimated that it would take another flight a minimum of 2 hours to get there even if they flew at maximum power and everything went smoothly. Bill and the remaining three planes would have to take turns shooting at the Japanese to keep them in the corral. This would eventually deplete their ammunition to a critical level, at which time the flight would be sitting ducks for a bunch of angry Zeros. Bill concluded that eagle red flight would have to do the job alone.

The training gunnery pattern was working well. The passenger cars all appeared to be empty, but he had an occasional ship strafe them to make sure no one was hiding inside or under the cars. Bill fired at the running troops in short 2- or 3-second bursts, and in order to disperse his stream of bullets, he walked the rudders and jacked the control stick back and forth lightly. One .50 caliber, with its 3200 ft/s speed, would incapacitate anyone it hit in a major member of the body. He noticed at this time that bodies were beginning to stack up like windrows in a hayfield. The rest of the troops were either in singles where they had fallen or spread-eagled tightly on the ground. Bill was suspicious of the spread-eagled ones. That's what he would have done if he were being shot at; either that or he would have tried to burrow under the stacks of bodies. He decided to concentrate the rest of his ammunition on the isolated troops and also hit the major stacks occasionally.

On his next pass, he decided to make a dry run to check out the cliffs to the east of the glen as well as the one leading down to the river. Sure enough, some of the braver souls were trying to climb down the cliffs. He called for Red 2 to hit the riverbank cliff and for Red 3 to shoot the ones trying to climb the cliffs to the east. Afterward, they continued to check the cliffs occasionally and to shoot the troops off them as required.

When they had been shooting for 15 minutes, Bill asked how much ammunition the others had remaining. He found that most of them were down to two or four guns still firing. He didn't fire on the next round and scanned the skies intently. It was about time for the Zeros to appear if they were going to come. So, he sent the third element back up to 5000 ft for top cover while the rest of them were firing out. On his next pass, he started getting a solid stream of tracers which warned him that he was down to 25 rounds per gun, or enough for one more good pass.

Bill decided to take a calculated risk and have all his birds empty their guns. This was too good a target to be conservative. When the lower two elements had fired out, he had the second element go up for top cover and called the other one down into the fray. As they were firing out, the wish for some napalm popped up again. This triggered a new thought: If they all dropped their almost empty belly tanks into the denser stacks of bodies, they might be able to ignite them with some .50-caliber tracers. They tried this. However, none of them were able to start a fire. "Oh, well," he thought. "The heavy tanks didn't do the troops much good." He asked his flight if anyone had any ideas on how to do more damage. Bless his heart, George said jokingly, "We have our .45s." "Eureka," said Bill, "here is what we are going to do. Put your gear and flaps down and slow the birds down to 125 mi/h, open your canopy, get down low, and have a try with your pistols."

They found that, with a little practice, they could get off two or three rounds per pass. They all felt like cowboys as they blazed away with their big automatics. They each fired out the magazine and reloaded. George reported once, "Hey, I shot my own wing tip." All of the flight came back on the radio in unison, "No, no, George. Aim higher or drop your wing lower." When they had fired their remaining ammunition, Bill considered having a go with his 6.35 Mauser but rejected this idea. It was designed for short range, and even at the 15 to 20 ft they were attacking from, it was too small to do much damage.

By then, they had been airborne for 1 hour and 40 minutes and they had been shooting for over 20 minutes. It was time to go home. They had fired 20,400 rounds of .50 caliber and 126 of .45 caliber. They scanned the sky for Japanese fighters and, seeing none, headed home.

On the way home, they discussed the number of passenger cars and how many troops they might have contained. The consensus was that there were 13 cars and that each held 80 to 120 soldiers. Using 100 troops as an average, they arrived at a total of 1300 troops. They were all convinced they had killed or severely injured 90–95 percent of the soldiers by the time they left. Red flight decided to officially claim 1100 dead and to report another 200 as casualties.

After landing and going through the ritual of debriefing with the squadron intelligence officer, Bill had a few words to say to them. He told them, "Guys, this was the most productive flight I have ever had and I suspect that this will be true for all of you also. You can all be proud for a job well done. But," he added, "I have a major criticism. I don't know who it was that I threatened to shoot out of the pattern and I guess I don't want to know either. However, if any of you ever quibble or refuse to obey a direct order when I am commanding a flight, you will never fly again. Are there any questions about this?" There were none.

That night, as they were sipping a beer in their makeshift club, two different flight members sidled up to Bill and asked him if he would have really shot the guy down. Bill's only reply was, "What do you think?"

The Introduction

Told by O. T. Ridley
First Lieutenant, U.S.A.A.F.

We had completed P-40 transition and training at Perry, Florida. It was mid-July 1944. Leave was granted to those who wanted to visit home. The rumor was that those not taking leave would go directly to fly P-51s in England. About 15 of the class decided not to take leave and hope for the P-51s.

Our group of 15 was processed for shipment and then given considerable instruction on how to board a train. This included exercises involving a mockup passenger train car. We filed in at a specific time (to the minute), proceeded down the aisle, and took our assigned seats (exact). Humorous comments were discouraged. The exercise was executed flawlessly. Having passed this hurdle, 2 days later we were allowed to board the real train. Destination was rumored to be Camp Kilmer, New Jersey.

July was very warm in the southeastern United States that year. We were crowded into one car together with some other troops (their rumored destination Camp Kilmer). The ventilation in the train car was poor, and when we opened windows, in came smoke, soot, and cinders. Sleeping was done in the seat, though a few hardy souls tried the floor. The trip was uneventful, except for an occasional lady that would wave or demonstrate other friendly greetings from trackside.

We arrived at Camp Kilmer after 2 days of travel, hot, tired, and dirty. The process of preparing for overseas shipment was not a lengthy one. We had the expected shots and examinations. We were issued appropriate clothing and gear, but most notable was the new B-4 bag. The B-4 bag was a remarkable piece of luggage, holding a large amount of clothing and other accouterments. It was the forerunner of today's light, soft luggage. The bag stood about 2 ft high and about 3 ft long. It had a good sturdy handle with some flexibility. The width of the bag varied with the load it was required to carry. It could be unzipped and would hang nicely on either side. The bag was made from a very substantial olive drab canvaslike material. There was always room for one more item in your B-4 bag.

The trip to England was uneventful. Quarters were below deck, with bunks stacked four high. It seems the man on the top bunk was always the one least accustomed to the rolling of the ship. The results of this improper placement usually caused the illness of all below the seasick man on the top bunk. We arrived in Liverpool in early August and were dispatched to the 496th Training Group, 55th Training Squadron (P-51s) at Goxhill, England. Goxhill was south of Grimsby and about 60 mi northwest of The Wash, on the east coast of England.

The 55th Squadron at Goxhill had P-51Cs. Here we were introduced to the Mustang. These airplanes seemed very light and responsive on the controls when compared with the P-40. The P-51 was faster and more maneuverable. It took some getting used to at this stage in our flying careers. We probably had an average total flying time of about 250 hours. The transition consisted of formation practice, familiarization with the countryside, and some mild aerobatics. We received about 10 hours flying time in the P-51. The transition took place without accident.

In early September, we were shipped down to the 357th Fighter Group located at Leiston, which is 30 mi south of Great Yarmouth, on the English east coast. We were greeted, assigned Quonset huts, allowed to pick our beds, and stored clothing and belongings. After dinner, one of the old boys in our hut turned on the radio and tuned to a German,

English-speaking news broadcast from across the Channel. "Lord Haw-Haw" was the news broadcaster. Two days earlier, the members of the 357th had been very active in West Germany. The Germans had been able to associate the group's red and yellow check nose colors with the 357th Group's airfield located in the vicinity of Yoxford, England. Thus, we were labeled Yoxford boys, with the threat of death if captured.

The next morning, we were scheduled to fly. The day was generally clear with good visibility. We were gathered in operations for the briefing. The airfield had three hard surfaced runways, each about 5000 ft long, with the usual taxiways and hardstands. We were assigned positions, call signs, and airplanes. The radio procedures were covered. The formation was to be one of general familiarization. The airplanes were P-51Ds. Special attention was given to the landing pattern. We were to come in on the deck, echelon right, pitch up and to the left, gear and flaps down, complete the turn, and land. The spacing on pitch up was 3 seconds between airplanes. The leader would take the left side of the runway; the wingman would take the right. We were to land in flights of four.

The two-ship formation takeoffs and join up went well. The leader, Captain John England, took us up to 10,000 ft where we did formation practice in the regular four-ship fingertip style—then echelon right, left, and then in trail. We were corrected as to proper position and timely movement. The formation was then placed in an eight-ship string, a few rolls were done, and then a few very tight pull ups and turns. Next Captain England took the string down across the airfield on the deck at about 300 mi/h. He then pulled up steeply into an Immelmann.

The first three airplanes in the string made the Immelmann turn. The fourth man, Lieutenant Jackson, lost control and went into a spin at about 5000 ft. I was not able to see the spin, but I did see the recovery. Jackson recovered over the maintenance hangar at about 100 ft. This scattered the formation like a shot into a covey of quail. The three airplanes left in the formation came in to land with echelon right. The leader, of course, had a nice tight pattern. The No. 2 man counted his 3 seconds and peeled up and to the left, gear and flaps down, keeping the leader well in sight and properly spaced. The No. 3 man made a go-around. Numbers 1 and 2 taxied to the hardstand. The rest of the formation straggled in with various patterns and landings. There were no accidents.

We were gathered as a group in operations. Captain England conducted the critique. The general formation flight was covered. He then got to the discussion of our aerobatics. We were the sorriest lot of pilots ever to be dumped on the 362d Squadron. He asked how many pilots were afraid of the P-51. No hands went up. At that stage, I suspect the only time we were coordinated was when the needle and the ball passed each other. Captain England then proceeded to tell us we were going to drive those P-51s down the main street of Berlin, and if we did not want to go, this was the time to get up and leave. There was a short pause while he waited for departures. There were none. Captain England then complimented Lieutenant Jackson on his spin recovery, thus ending the critique. I was blessed during this exercise by being the No. 2 man in the formation.

Another Interesting Day

Told by Roy D. Simmons, Jr.
Captain, U.S.A.A.F.

August 1944, flying from Borgo Airfield, located south of Bastia, Corsica; my airplane *Flying Jenny* an F-6C (P-51C) assigned mission to accompany a fighter bomber squadron from the 86th Fighter Group (or was it the 79th?) to the Po Valley in Italy, observe and photograph the destruction of a railroad yard and bridge, and then check the airfield at Pisa, Italy, on my return to Corsica. This didn't seem very difficult; however, things are not always as they appear.

The mission briefing officer stated that XII Tactical Air Command was quite anxious to receive the results of this mission; therefore, I was to return posthaste. I was assigned as my wingman one of our new pilots. He turned out to be quite competent, eager, and someone who could follow instructions. As we took off and as the wheels were retracting into the wheel wells, I glanced to my right and there he was tucked in just like a professional. The mission was off to a good start.

The fighter bomber squadron we were to accompany was stationed south of us; therefore, our takeoffs were coordinated, which enabled us to rendezvous over the Mediterranean. A beautiful clear morning made rendezvousing simple, as I placed our aircraft to the right of the squadron and proceeded to the target area. I'm not so sure we were appreciated, tagging along, but we also had a mission to fulfill.

As we neared the target area, the fighters positioned themselves for their bombing and strafing runs. Meanwhile, we remained clear of the target area, observing and waiting for the bombing and strafing to cease and for the fighters to leave the area. Now it was my turn. I positioned my wingman away from the immediate area of the target, instructing him to keep me in sight and to be on the lookout for enemy fighters as I dove to 1500 ft. There was no need to further endanger him to the flak, which was extremely heavy. It was obvious the fighters had stirred up a hornet's nest. It seemed that everything was being thrown into the air, including the proverbial kitchen sink. I only hoped and prayed that none of this flak had my name inscribed on it.

Getting into position, I began my dive, calling on *Flying Jenny* to hold together and give me all she had. Leveling out at 1500 ft, throttle wide open, remembering to fly straight and level, maintaining a constant heading and altitude, I was all the time flying through thick black smoke and bursting flak. If a combat pilot says he was never scared, he's lying. This one admits to being scared. After making the photo run, I dropped down to treetop level; to pull up immediately would have presented a beautiful target for the ack-ack gunners. While at treetop level, I saw three trucks in my flight path, so as any good fighter pilot would do, I gave them a short burst as I flew over. Looking back, I saw two on fire. Pulling up, I sensed this part of the mission had gone quite well. I called my wingman to join me. A good feeling came over me as he slid under my right wing, giving me a big smile and a "thumbs up."

The flight to Pisa was an uneventful trip of about 100 mi. The Leaning Tower of Pisa always fascinated me; however, as we were getting closer and could see the tower, it held little interest for me that day. Another hostile beehive to enter, as evidenced by the flak welcoming committee. Should this be a visual or photo run? It was my decision. I chose a photo run, as there seemed to be considerable activity on the airfield. Again, I positioned my wingman away from the target area with the same instructions as before: Stay clear, watch me, and observe the area for enemy aircraft.

I started my dive to 1500 ft, leveled out, making my run with intense flak getting closer. All of a sudden, I felt *Flying Jenny* shudder. I'd been hit! Looking out, I saw a gaping hole in the right wing. "Stay on the run," I kept saying to myself, for to pull up would have given the ack-ack gunners a beautiful target. "Stay low, stay low until out over the Mediterranean," a voice kept saying. After what seemed like ages, I started a slow climb, right wing seemed heavy, caused by drag, and called my wingman to inspect the damage. He quickly joined me, slid beneath, and reported no visible signs of oil or hydraulic fluids streaming from the airplane. I then slowed down and checked the handling characteristics, noticed a slight change, a little more trim needed to keep the right wing up. By now, I could see Elba Island on my left and the faint outline of Corsica straight ahead, so I headed home. As I neared the airfield, I notified the control tower of my problem and had them clear the runway for me.

Mission accomplished, and yes, the fighter bomber squadron demolished the railroad yard, and the bridge was of little future use. So ended another successful mission. *Flying Jenny* was repaired and back in business in a few days. (The Leaning Tower of Pisa still fascinates me, and 18 years later, in 1962, I had the thrill of climbing to the top of it.)

Mortain and the Oracle at Bletchley

Told by Joe Thompson, Jr.
Major, U.S.A.A.F.

It was 10:10 a.m. on August 3, 1988. I stood frustrated in the hot, littered attic. How could one family accumulate so much stuff even if we had run four children through this house 30 years this November? Then I saw the tattered, corrugated paper roll, 6 in across and over 4 ft long in the far corner of the chaotic room in the attic. I grabbed it. The lost was found. Here was a complete set of combat maps of Europe, World War II. By 10:20 a.m., I was on my knees in the living room searching for the two sections I needed to use once again. They were there and 44 years came floating back as yesterday but without the fear, the trauma, and yet with the high adventure of those crucial days.

June 4, 1944. A solid overcast at 3500 ft as we left Middle Wallop and flew past the coast, crossing over above Bournemouth, with a compass heading for Le Havre. We flew at 50 ft above the Channel, hoping the wind drift was not greater than we had calculated and keeping low so as to escape the German radar until the last minute. Landfall a bit northeast of Le Havre, with very light flak and then we were into the French countryside. The sullen skies were lower now, but patchwork forests spread out below us like a homemade quilt. Northeast of the Seine between Rouen and Paris, the woods were so distinctive and widely spaced that you could navigate by their shapes without a map. The missions called for one no ball target in that area and then two or three airdromes west of the Seine between Paris and Argenton and other points further west.

No ball targets were the launching platforms for the V-1 planes, those pilotless flying bombs with a set compass heading and a timer to cut off as the planes reached the London area. The Germans always built the platforms inside a wooded area, and the runway looked like a big hockey stick with the handle pointed toward London. We photographed them. The medium bombers hit them. Then we went back to see whether they'd had a strike or a spare.

"Peafowl Yellow 1, flak at 8:00." "Roger." My No. 2 was to call all flak and enemy fighters. We flew by the maxim of, "He who flies and runs away will live to fly another day." We went on to check out an airdrome at Evereux and then to Conches. In April and May, both the dromes had FW-190 squadrons. Now they appeared inactive and almost deserted. The same was true of the airdrome at Dreux, somewhat southwest. It is as if no one could scramble fast enough to catch you. And at 3000 ft, those six-gun flak batteries could strike without warning. We made it past all airdromes without incident. No rail traffic, no truck, troop, or tank traffic sighted. Peaceful enough thus far. From Vire to central Normandy, we headed to the last target. It was a prominence on the Normandy coast not far from Bayeux. We had photographed the whole coast a couple of months ago from an altitude of 50 ft, but this was a vertical at 2500. The clouds were lower still and no calls from Yellow 2. I got the photo and "put everything in the kitchen" heading home. One hundred and forty miles of Channel and we were safe in England. The prints were delivered to Eisenhower's headquarters at Wilton House in Salisbury that night, and 2 days later I learned the last target overlooked Omaha Beach.

How minuscule were the pieces of the puzzle given to each unit or squadron and how little we knew of the gigantic influences that shaped the destiny of the war and caused us to receive orders that were sometimes horrendous, always hazardous, but in a few instances, provided a front row center seat with no one taking up tickets.

Hitler, the infallible military genius, signaled Kluge, who had been placed in charge of the Normandy front in place of Von Rundstedt, that he had taken over the whole Western theater. Then, on August 2, came the most important of all German signals intercepted during the war. Hitler said, "Pay no attention to the American breakout. Collect four armor divisions with supporting infantry. Retake Avranches and push the Americans into the sea."

For the first time in the history of the German responses to Hitler, Kluge, a methodical and able German general, had the audacity to respond to his Fuhrer that such action would be near suicidal. His long explanation of his viewpoint was provided in its entirety by the Oracle wherein he stated that essential divisions were needed to hold Caen, but if such an attack were not successful, it would lay open the whole attacking force to be cut off in the west. Poor Kluge. He had just been given command of the German forces, and now he saw an impossible duty forming. The next day, Hitler responded, acknowledging Kluge's arguments, but stating, "The situation demands bold action. The attack to split the American forces must be carried out." Because of the length of time it took for these back and forth arguments, 3 days had passed, giving Bradley, Eisenhower, and Patton just the time they needed to prepare for the attack that Hitler had directed.

However, the stubborn Kluge wasn't finished yet. On August 5, he had one more try to dissuade Hitler, but to no avail. Kluge was staking his whole career on trying to stop the idea of this attack and in his last signal stated it could only end in disaster. The Fuhrer replied without comment; the order was to proceed as instructed.

Back on A-9, the missions posted for August 8 requested Peafowl Yellow Section to report in detail the enemy action in Section 6, covering the lower part of Normandy, and included

Avranches and Mortain. Of course, we knew nothing of the German counterattack, only that Patton had made it around end at the base of Normandy and was about to head into clear running toward and beyond Paris. We took the route west of St. Lo, now a shambles with no building standing, and quickly reached Villedieu les Poelles, a small town on the rail line between Granville and Vire. Fifteen miles south would be Mortain, below Sourdeval. The road from Mortain to Barrenton had some wood patches, and down we went for a close look. Here was the front row center seat I had been promised. Hitler's armored columns for the counterattack on Mortain clogged the road. Tiger tanks with camouflaged netting were moving slowly on this narrow route. We were down to 100 ft. I could see the German soldiers clamoring into their open jump seats as we roared overhead. So quick was our appearance and departure that not a shot was fired either by the Nazis or by the two of us.

I pulled up sharply, zigzagging furiously to avoid the flak I was sure would follow. Then I saw, perhaps a mile away, a flight of P-47 fighter bombers. While we did not have the same radio frequency, the age-old wing waggle was sufficient to attract their attention. At first, it looked as though they would not get the message, but then they followed us down to the location of the attacking column. They dove with machine guns blazing, 250-lb bombs under each wing. "Peafowl Yellow 1, flak at three o'clock." "Roger." "Out." We headed back to A-9, little knowing we had witnessed the end of the Nazi penetration, and the closing of the Falaise pocket, while not quite completed, had so dispersed the German armies in the west that all of France would now be cleared and the end of World War II was in sight.

The Ultra secret, the Oracle at Bletchley, had indeed destroyed the Hitlerian magic, and the thousand years of rule by his so-called superior race were negated. There were, of course, pieces of this puzzle still missing, and the power of his German war machine still had a few sparks left (we were to fight the Battle of the Bulge without help from Ultra, since Hitler had ordered all signals to cease). But our squadron had the dubious honor of continuing to give flight cover to the First Army through France, Belgium, the Rhine Valley, and finally central Germany. We would never have reached those points, however, without the resounding defeat of the German armies in the Battle of Normandy. The Oracle of Bletchley was the difference.

Project Anvil

Told by William W. Wells
Captain, U.S.A.A.F.

It was one of those rare summer days in England, warm and sunny, August 12, 1944. On this day, the 20th Fighter Group flew two main missions. The first one was east of the Seine River escorting B-17s on another bombing mission with our P-51s strafing their way back home, hitting any moving form of transportation. The second mission, just east of Paris, was similar to the first; bombing by heavies, fighter bombing, and then strafing targets of opportunity. Both were quite successful.

In the meantime, those pilots not on the main missions were on standby in their respective operations areas. About 1600, the phone in the 55th Squadron operations rang. Someone from group operations called to tell us to get a flight of four P-51s in the air at once and fly to a field called Winfarthing-Fersfield located about 15 mi southwest of Norwich and about 60 mi east of Kings Cliffe. We were to report to a Commander Smith of the U.S. Navy. Lieutenant John Klink and I, with two other pilots (unknown at this time), took off as directed and reported to Commander Smith at Fersfield for the mission briefing.

Commander Smith took us into the briefing room and closed the door. He began by saying this mission was, next to the invasion of Normandy itself, the highest classified mission of the war. In fact, it was 17 years after the war before the facts of this mission, called Project Anvil, were made public. Needless to say, he swore all of us to secrecy.

Commander Smith told us of the extreme concern of the Allied high command over the unleashing of the German V-1 and V-2 robot bombs. For months, the Allies had made countless bombing strikes on suspected V-1 and V-2 launching sites without much success and yet with the loss of nearly 450 planes and 2900 aviators. Some other means of stopping this onslaught was desperately needed.

Commander Smith explained how this Project Anvil was to be implemented. He pointed out a PB4Y Liberator (B-24) painted all white parked on the side of the field. This, he said, was our drone, but because of its heavy weight, it would require a pilot and copilot to take it off and get it set on course. After it was on course and before it reached the Channel, the two pilots would bail out, leaving the PB4Y under control of the accompanying radio-control planes.

He went on to tell us the drone was loaded with 374 boxes containing 55 lb of Torpex each and six Mark-9 demolition charges containing 100 lb of TNT each. This came to 21,170 lb of high explosives. He then told us of other planes which were to participate in this project. The first plane to take off was an RP-38 photo ship, next a Mosquito photo recon ship, and then two PV-1 Venturas especially equipped with radio equipment for guiding the drone. After this, the drone (PB4Y) itself would take off and finally the four P-51s. We were told to climb up and maintain about 1500 ft above the PB4Y drone at all times until it was launched onto the target. We were then to accompany the PV-1s back to Fersfield base and report to Commander Smith.

At 1752, the aircraft proceeded to take off in their proper sequence. The drone leveled off at 2000 ft, the photo planes went up to 15,000 ft, and the PV-1s were also at 2000 ft but about 2000 ft on either side of the drone. We took our position at 3500 ft and throttled way back, constantly turning so as not to overrun the drone.

The Nazi-occupied island of Helgoland was the target since this island was a proving ground and launch site for V-2 missiles. I'm sure all the Allied pilots remember this island fortress in the North Sea. We drew flak from here every time we crossed the Danish coast.

The remote control systems of the PV-1s and the drone were to be checked at five points. The first point, "Able," was over Framlingham about 25 mi southeast of Fersfield. Next the drone would turn north to point "Baker" at Beccles, then south to point "Charlie" at Clacton-on-Sea, and on to point "X ray" near Manston Airfield where the PV-1s would take control and the two pilots would bail out. Then the drone would be directed to point "Dog," about 2 mi south of Dover, and thence toward the target some 350 mi to the northeast.

The drone, its control ships, and the P-51s moved toward checkpoint "Able." Just before they reached Framlingham, the drone pilot signaled by radio he was ready to conduct the first radio-control check. Point "Able" was passed at 1815. The group then turned left for point "Baker" over Beccles. We flew over Heveningham Hall, the home of Lord Huntingfield, crossed the Blyth River, and could see the North Sea off our right wings. The tall tower of St. Michaels loomed ahead in the town of Beccles. At 1820, the drone was over a field near the villages of St. Margaret and St. Lawrence. We, in the escort P-51s, were in a slow, lazy right turn and were looking right down on the drone directly below us. Suddenly, it exploded in a searing orange ball of fire, blowing us up into the air an extra 1000 ft. A huge mushroom cloud boiled up to 25,000 ft, and nothing remained of the drone or the two pilots.

We returned to Fersfield and landed. We told Commander Smith what had happened even though he could hear the blast and see the high cloud. He told us, sadly enough, the same thing had happened the previous week to another drone. He also told us the pilot, in this joint Army-Navy Project Anvil, who had been on our mission was Joseph P. Kennedy, Jr., the son of the Ambassador to the Court of St. James, Joseph P. Kennedy, Sr., from Boston. Young Joe was an outstanding flier and had volunteered for this mission knowing full well how dangerous it was. A true hero, he was awarded the Navy Cross posthumously.

On Sunday morning, August 27, 1944, approximately 1200 heavy bombers took off from English airfields escorted by some 1000 fighters headed for Berlin. After we crossed into France, we learned the weather around Berlin was exceptionally bad, so we were advised the mission to Berlin was canceled. We were also told we were free to attack targets of opportunity in West Germany, Denmark, and northern France.

Some of the bombers elected to bomb several factories known to be manufacturing trucks, aircraft, and aircraft engines. Other targets were railroad stations, docks, storage depots, and ammunition plants. The weather was clear in northern France and western Denmark, so we concentrated our efforts in this area.

Since no German fighters appeared that day, we had the run of the entire area. Our 36 bombers decided to hit the air base in Esbjerg on the Danish coast. They told us on the radio that as soon as they finished their bomb run on the airfield they would let us know. In the meantime, we dropped down and shot up trucks, trains, cars, power stations, and especially anything on wheels.

Soon we saw the columns of smoke and dust over the airfield at Esbjerg and the bombers called to say all was clear for us to come on in. We had climbed up to about 12,000 ft. There were 20 of us in P-51s while the rest of our group gave us top cover. We let down sharply at full power and headed for the airfield with gun switches on. We picked up speed to about 425 mi/h and dropped down to about 50 ft above the ground. As we came over the edge of the field, we pulled up to about 100 ft to get an angle for firing at the field's targets, whatever they were. All 20 of us opened fire at anything that looked like a target—planes, gun emplacements, trucks, buildings, and hangars.

All of a sudden, groundfire opened up from all sides, the most intense antiaircraft fire any of us had ever seen. Before we knew what was happening, four of our planes were going down. One crashed into the North Sea and the pilot was killed. Three others managed to bail out and ended up as prisoners of war. The rest of us were shot up but managed to head for home some 350 mi across water.

I looked at my engine and saw that two large caliber shells had exited on the left side. I anxiously waited for the gauges to tell me I was in trouble. I expected the engine to quit any second. But nothing happened. I looked around for my teammates, found several, and soon 16 of us were in formation. We lost our commanding officer, Lieutenant Colonel Wilson, and three other good pilots.

When we arrived at our home base, we discovered that everyone had taken some hits. In addition to two through my engine, I had four just behind my canopy and six more holes in my tail section.

We didn't realize the magnitude of this mission until later, but it was the largest and most devastating attack in Denmark during World War II. We only played a small part that day, but remember, there were 1200 bombers and nearly 1000 fighters all attacking their targets of opportunity at the same time, mostly in Denmark. Since this day was so shocking and traumatic for Denmark, two Danes have written a book entitled *Luftangreb Pa Vestjylland, Aerial Attack on West Jutland* (Denmark). It is a compilation of eyewitness accounts of action that day, and those of us in the 20th Fighter Group were featured, with photographs.

It is sad now to read eyewitness accounts of the death and suffering of hundreds of Danish civilians. However, many of these people were either willing or unwilling collaborators and were simply victims of a terrible war. If the weather in Berlin had been better, it may not have ever happened.

Not So Routine

Told by Roy D. Simmons, Jr.
Captain, U.S.A.A.F.

August 22, 1944, flying from Borgo, Corsica, into southern France, 84 combat missions completed in 4.5 months, wingman with 11 completed combat missions. Briefing consisted of close surveillance of Germans pulling out of southern France, moving north up the Rhone Valley. It was to be a long mission while sitting on an unforgiving, hard dinghy. Lieutenant Hoy, my wingman, was eager to get to France, observe the retreating Germans, and complete another successful mission.

After crossing the beach area west of Cannes, we became aware of retreating Germans and the absence of flak. The closer we got to the Rhone Valley, the greater the volume of enemy movement. Photographing this vast movement was essential.

Well, look what I see. There goes a train heading north, one of those "juicy" targets you could not resist. So with Lieutenant Hoy in tow, we headed for the train. Strikes on the train were beautiful, seeing it slow down with a white substance coming from the engine, which apparently was steam. We made two passes on the train and departed as the flak was becoming intense. We felt good as we left the area and continued our mission. Gad, that dinghy was hard.

It wasn't long before we came upon a few straggling trucks from a convoy. Remembering our mission of determining the extent of the enemy retreat, I thought better than to go after the trucks. We hadn't gone far before I could tell that Lieutenant Hoy wanted to hit the trucks. Why not? We

172

dropped down and began strafing the trucks. I saw the one I was firing at burst into flames. Lieutenant Hoy was so intent on destroying his target that he flew over a truck as it blew up, debris hitting him, causing him to crash. No possible chance of survival. I pulled up, gained some altitude, and circled the area. There was nothing I could do and the flak was getting heavy. I completed the mission and headed out over the Mediterranean toward Corsica.

A lonely feeling came over me. Here I had been deep and alone over enemy territory, lost my wingman, getting late as the sun had already gone down, and to cap that off, my engine was acting up. It became so rough I took a survey of my surroundings, loosened my seat belt, and checked the canopy release. I looked down at the vast amount of water, preparing to leap out if the engine should fail. I kept nursing the engine because in the distance I could make out the coastline of Corsica and that meant "home."

Calling the tower, a lone airplane inbound, engine trouble, requesting a straight in approach, getting a "Roger." What a wonderful feeling when the wheels of *Flying Jenny* touched the runway.

Debriefing was tough, mission accomplished, my 85th mission a success, but why did I not feel good about the results? Tomorrow, another day. Yes, I'll be flying No. 86. Just another day in this crazy war.

European Theater of Operations (Southern France): Combat Mission 144

Told by Herman K. Freeman
Lieutenant Colonel, U.S.A.A.F.

The Allies established a beachhead in southern France in mid-August 1944, and the war was moving very rapidly to the north. My squadron was flying out of Corsica in support of the invasion, and we knew that as soon as the engineers could get a landing strip ready we would move onto the beachhead. About 10 days after the invasion, we got orders to move the entire group into France and we moved on August 25, 1944.

The war was moving so fast and the front lines were so indefinite that about all we could do was fly road reconnaissance missions at a safe distance ahead of the front lines so we would not endanger our own troops. On September 4, 1944, I was ordered to lead a flight of four P-47s on a road recon mission in the area to the west of the city of Dijon, France. We were not carrying any bombs, as we were carrying belly tanks with fuel so we could patrol longer. Each plane was armed with eight .50-caliber machine guns, which can destroy about anything we would run into. We were to strafe anything that moved, as we were trying to prevent the Germans from moving troops or supplies into the battle zone.

After patrolling for a short time, we spotted four vehicles moving south toward the front, so we headed down for our strafing run. We destroyed them in one pass and got two of them burning—we call them "flamers" in fighter pilot lingo—and the other two were badly damaged. We then climbed back to our patrol altitude to look for another target.

About 15 minutes later, we saw a target that every fighter pilot dreams of. We were on a westerly heading and I spotted a train off my right wing also heading west. I called the flight and told them we would make our strafing pass from the west so we could strafe the entire train from the engine back. As I started my run, I identified the train as passenger cars. The engineer must have spotted us and he got the train stopped right between two large open fields. The German troops on board were coming out all the doors and windows in one big mad scramble. The first ones out dove into the ditch along the tracks, and when that got full, they just ran across the field in an effort to get away.

On my pass, I hit the engine first and got the big billow of steam from the boiler being punctured and then I kicked a little rudder to put my sight on the ditch that was full of troops. There were so many troops running across those fields that it looked like a football stadium emptying after the big game. I told the rest of the flight to concentrate on the troops, as I had already put the engine out of commission. I pulled up off the target and told the flight to join up, as I figured we had done as much damage as we needed and it isn't too smart to hit the same target right away because you don't know what you will run into on the second pass.

We couldn't have made a second pass anyway as I got a call from my No. 4 man who said that he had hit a tree. I looked for him and saw him climbing to join up, but he seemed to be having a little control problem. I throttled back so he could join up, and as he got closer, I could see his problem. His left wing was gone from the inboard aileron hinge. The aileron was still attached to the wing by that one hinge and it was just trailing in the slipstream. He seemed to be able to control the plane all right, but he told me he couldn't see anything through the windshield as it was plastered with green from the tree leaves. I told him I would lead him home and we would land in formation, but if the aircraft became uncontrollable, to be ready to bail out. We were about 125 mi north of the base and flying at about 4000 ft, so I knew we had a good chance of getting him home okay.

I advised the tower that we were coming in on an emergency and to have the runway cleared for a straight in approach. I got an affirmative and started a gentle letdown about 10 mi out. I told No. 4 to put his gear down, but if he started having any additional difficulty with control, to be ready to retract it. He put the gear down and everything worked normally with no problem. I kept our airspeed about 20 kn above normal on approach because I didn't know when that left wing would stall. As we were approaching the runway about a half mile out, I saw a fuel truck approaching the runway on the perimeter road en route to the fuel dump. I was sure he would stop and hold short of the runway, but he evidently didn't see us and drove right out in front of us. I hollered to No. 4 to stay with me and pulled back on the stick just enough to clear the truck and proceeded with the landing. We made it all right and got stopped in plenty of time.

But I was so mad that I parked my plane and went looking for the fuel truck driver. Everyone I checked with could see I was a little out of control, so none of them admitted knowing who the driver was. He was being protected by all his buddies, and looking back on the incident, I'm glad they did because it kept me from making a complete fool of myself.

On our critique with the intelligence officer, we made a conservative estimate of 200 to 400 casualties. We lost the use of one of our aircraft that could not be repaired, but we all got back safely, which overall made it an extremely successful mission.

Severe Weather

Told by William Shwab
First Lieutenant, U.S.A.A.F.

In September 1944, the 311th Fighter Squadron, along with the entire 58th Fighter Group, was transferred from New Guinea to Noemfoor Island, a small coral atoll about 100 mi northwest of Hollandia, New Guinea.

On this mission on October 20, 1944, we had a 12-ship flight from Noemfoor to Ambon, near Borneo. This was to be another long 5- to 6-hour flight, mostly over water. As a result, we carried two external wing tanks, each of which carried about 100 gal of fuel to give us the additional range needed. The only other "external" was a 100-lb bomb on the belly.

On hitting the target at Ambon, we used our bombs on a fuel storage plant that one of us set ablaze. Then we strafed the revetment area until we had used up most of our firepower. We then re-formed on our flight leader, Lieutenant Max Itzkowitz, and started our return to our home base at Noemfoor. We had now been in our airplanes 3 hours, and as we started back, I could not help but notice the severe weather. We were trying frantically to get in front of the weather moving in, and our flight leader asked how much fuel we had left since his main tank was getting low.

All of us in the flight had just enough to get home, but we flew under our leader and noticed he had been hit and was losing fuel. Our element leader, Tex Kindred, and I advised Lieutenant Itzkowitz that we would cover him as we looked for an island shoreline for him to ditch on. This he agreed to, and in a few minutes, we were able to find a small enemy territory island for him to belly in on, wheels up on the coral island.

We were able to contact air-sea rescue based out of Biak Island, just northwest of New Guinea. They were using Navy PBY Catalina flying boats that patrolled this large area over enemy territory, and one of them was able to rescue Lieutenant Itzkowitz. He had sustained an injury to his arm but was returned safely to base the next day. Several others in our squadron ran into the severe thunderstorms and had to make emergency gear-up landings in the ocean or on enemy-held islands. Of the eight planes that went down, at least five of the pilots were picked up by the two "Cats" that were in the area.

The three of us left in our flight were just barely able to make it back due to shortage of fuel. We had completed a 6-hour flight, which was the longest combat mission I experienced.

Operation Market Garden

Told by Enoch B. Stevenson, Jr.
Major, U.S.A.A.F.

Sunday morning, September 17, 1944, was an absolutely perfect day. The sky was blue, the rain had moved out, and the wind was fairly calm. Our squadron, the 503d of the 339th Fighter Group, took off at 0830, formed up, and set course for the vicinity of Arnhem, a lovely and affluent city in northeast Holland, very near the German border.

The weather worsened over the Continent, but we were able to operate underneath the overcast. We found no targets of significance, but we did strafe some trains and other transport. Everyone returned to our base at Fowlmere, Hertfordshire, at 1245 with no casualties and no battle damage.

Little did we know that we had preceded the largest and most ambitious airborne operation in history. At 0945, and for 2.5 hours thereafter, some 2000 troop-carrying planes, gliders, and their tugs departed 24 U.S. and British bases. They created an awesome sight when we saw them as we returned over the North Sea. Miles and miles they stretched. Their formation flying was not exemplary, but they were going in the same direction. We returned to Fowlmere without incident.

All of this activity was preceded by an attack on flak and artillery positions by some 1400 heavy bombers, U.S. B-17s and B-24s. Apparently, the results were quite destructive. The entire procession of some 4700 troop carriers, bombers, and fighters constituted the largest airborne attack ever mounted. The order of departure was the British First Airborne Division commanded by Major General Urquhardt, followed immediately by the U.S. 82d Airborne Division commanded by Major General James Gavin, and the U.S. 101st Airborne Division commanded by Major General Maxwell Taylor. Despite terrible weather over the target area, the drop was successful, especially in the areas of the 82d and 101st Divisions.

Later, the weather soured considerably in England and on the Continent to such an extent that air operations were inhibited for several days, resulting in disaster for the British First Airborne. They fought gallantly, but to no avail. The bridge at Arnhem was never secured for the advancing forces, delaying the liberation of the Netherlands for some 6 to 9 months.

Beach Workout

Told by W. S. Whitmore
First Lieutenant, U.S. Marine Corps

My wingman and I were looking for targets of opportunity in the Marshall Islands when we caught a single soldier out on a coral reef trying to get to safety on the nearest islet. We began a series of strafing runs with rapid turnarounds to get him before he made it. On each 180, we got slower and lower, and my wingman finally got so slow that his plane began to settle in a stall. I watched as he added full power in a nose high attitude just as his wings stalled, but the 2000-hp engine pulled him out of the water and we called it quits for the day. The little unarmed Japanese soldier made it to the trees.

I spotted two Japanese spearfishing on a reef and began a strafing run hitting one who went down in the shallow water while the other made it to shore. I watched as the man turned around to help his injured friend, and as I passed over, I saw the bloody water and spotted a huge shark swimming toward the man. I wanted to turn around to get the shark but had to continue on. I am not proud of that mission.

My wingman and I spotted a group of about eight Japanese on another target of opportunity mission in the Palaus. They were on a small beach that ringed a large rock off the main island of Babelthaup. They were all in swimming trunks, and seven of them made it into a small cave or hole in the rock. There was no room for the last man, and he ran at full speed to the north side of the rock as we began to strafe the beach. We took turns strafing with the little No. 8 running from one end to the other. We shut down four of our six machine guns to conserve ammunition and to cut down on the sand which obliterated our target. We had made about a dozen runs when my friend finally leveled off just above the water and began his last strafing run. His shots bounced back 180°, severed an oil line, and he barely got back to Peleliu with smoke pouring out of the engine. We never did get No. 8, but we sure gave him a good workout.

European Theater of Operations (Southern France): Combat Mission 145

Told by Herman K. Freeman
Lieutenant Colonel, U.S.A.A.F.

The mission I flew on September 8, 1944, is a good example of the dire consequences that may occur if orders are ignored or not followed. Our group had just moved from the southern France beachhead area to a new base near the front lines at a town called Amberieu. It was situated in France just northeast of Lyon and southwest of the western end of Switzerland.

My first mission out of the new base was to lead a flight of four P-47s on an armed reconnaissance flight in the area where the boundaries of Germany, France, and Switzerland meet. It was known to us as the Belfort Gap. Shortly after we arrived in our designated patrol area, we spotted a train heading west from Germany. It was a bit unusual as the engine was putting out quite a bit of smoke. The train was just a mile or so short of crossing a high trestle across a river, so I told the flight we would hit it while it was on the trestle. I started my run from about 4000 ft and planned on getting down in the riverbed so I would almost be on a level with the train while I was strafing.

As I was descending in my dive, I saw the train come to a stop on the trestle, and the whole train seemed to sparkle. I realized immediately that we had been suckered into attacking a flak train of about 15 flat cars. I broke hard to the left while calling the flight to break off and don't go in as it was a flak train. My wingman followed me, but my element leader and his wingman proceeded with the attack. When they came off the target, my element leader said he had been hit several times but was still flying. His wingman was smoking badly, and I tried to contact him but could not get through, so his radio must have been out.

I told my wingman to escort my element leader back to base and I would stay with my No. 4 man. I could tell he wasn't going to make it as the engine was smoking so badly, but he was too low to bail out. His only choice was to belly in, so he headed for the only open area in sight for a wheels-up landing. It looked like a pasture sparsely dotted with large trees. He made a good belly landing, but the aircraft hit a large tree with the right wing and sheared it off right next to the fuselage. I buzzed the crash and could see the pilot in the cockpit, and he was slumped over in his seat and not moving. I was being shot at from some gun positions in the woods but couldn't spot them. I made two more passes over the crash, but the pilot hadn't moved at all and I was being shot at all the time so departed the area.

When I got back to the field, I had a meeting with my element leader and my wingman. My wingman said he heard my transmission loud and clear to break off the target. My element leader said he didn't hear me say anything. I didn't believe that because he had Rogered my transmission of hitting the train on the trestle. I told him that even if he didn't hear me, he was supposed to follow me when I broke off the attack. He was lucky to get home, as he had been hit four times with 20-mm explosives: one in each horizontal stabilizer, one in his left wing, and one in the top of the engine. The engine had lost all the oil and the aircraft was covered with it. When he landed and the aircraft slowed down, the engine overheated and froze immediately. He said when he went over the train, he had been flipped on his back, and it was only luck that he didn't crash then.

I didn't officially file a protest on my element leader's disobedience of orders because we had a war to fight and there was no way to prove it. In fact, maybe he heard me, but what I was saying didn't register in the excitement of the moment. Even so, I will never forgive him for costing a young pilot his life.

A "Tallyho" and a "Hey, Rube"

Told by Kenneth West
Lieutenant Junior Grade, U.S.N.R.

October 27, 1944. My logbook says combat air patrol and it isn't even noted for the combat engagement. The book was close but the author wasn't there, nor was the person telling the story.

The last time I saw the other person involved was on November 11, when he was pulled out of the No. 2 standby spot, gave me a palms up "what the hell," and was launched in my place. He never returned. "Swede" Thompson was generally my section leader. You know what I mean. He not only outranked me, my name starts with a W. Tail End Charlie most of the time.

So back to my story. It was kind of a combat air patrol. It was during the Battle of the Leyte Gulf and our Air Group 15 was very much involved. We were looking for the Japanese fleet, which involved sector searches of 300 mi across the Philippines. Due to the enemy aircraft that might be encountered, each sector was flown by a four-plane fighter group. Since communications might be garbled due to distance, two sections of two planes were sent out to relay messages. The first was put 100 mi at 5000 ft and the other was out 200 mi at 10,000 ft. Swede and I drew the second spot, 2 hours on station above Marinduke Island, which was known to have a Japanese airfield.

My comment to Swede prior to takeoff was, "It's a hell of a note to send out a two-plane fighter sweep." It turned out to be a bit more than that. We were in F6F5s just received and our armament in addition to the six .50s was six wing-mounted 5 rockets. It was the first time we had ever carried them and they armed a bit differently than the .50s.

We arrived on station with me sitting in the right wing position. We had made one 360 orbit on station when I spotted them. They were single-engine bombers and fighter bombers made up of three groups of three 9-plane sections on our reciprocal course. One hell of a lot of aircraft.

I gave a "tallyho," many bogies, two o'clock angels 7, and a "Hey, Rube," and we peeled off for a high side attack. I armed my rockets but unknowingly immediately turned them off. Instead of locking them in place by flipping the switch cover back down, they had to stay up. Just the opposite of the .50s.

Our attack was a complete surprise. Certainly they weren't expecting any and luckily we were coming out of the sun. They didn't know until we hit them that we were there, and they scattered to the four winds plus up and down. We pulled up in a high wingover and came back across on the second pass before they had any chance to organize anything. On this one, I rearmed my rockets and jettisoned them. We didn't need any extra baggage. We had already gotten rid of our belly tanks.

On our third and last pass, I noted two or three splashes in the water; however, it could have been rockets. As we were peaking out after this run, a Zero zeroed in on me and all of a sudden my windshield got an oil bath. I yelled, "Let's get the hell out of here," and we headed for daylight with several fighters following us. Thinking I had been hit, Swede tried to drop back to protect me, but I slowed with him and moved into the thatch weave position. We were going balls out and added a little water injection for a brief time. Most of the planes chasing us dropped out, but one was persistent. He finally made a pass at me, but defensive action sent him home.

For some reason, one of the things that concerned me most was being over land. If my engine quit, I would have to bail out. Wouldn't have even thought about it had I been over water.

After the chase, we headed for home. Swede couldn't home in and we were way off course somewhere. I took the lead, guessed at our position based on our estimated course and speed while being chased, checked the whitecaps for our estimated direction and wind velocity, and plotted a course to take us home. It worked just like it said in the book. Missed the disposition by about a mile and the ETA by about 3 minutes.

I landed aboard the *Lexington* because the *Essex* wasn't able to take me aboard. It's the only wave-off I got while operational in combat. Our first officer, Roy Bruninghaus, would have given me a high and fast cut and I would have been aboard. They fixed up my plane and I was back aboard the *Essex* the same afternoon. With a line or two shot out, the engine had lost only 16 of 19 gal of oil on the trip home. That engine was a good piece of equipment.

We never did know how many planes we got. I figured Swede got three and I maybe two. Word got back to the *Essex*, five planes, and they thought I had that many. With so many planes in the air, there ain't no time to watch them splash! Shoot. It's hard enough just to stay out of their way.

We got recognition for breaking up an enemy strike force headed for the disposition. Of course, we could have been court-martialed for leaving our station as a communications link messenger for the long-range searches. Should there have been a contact, that surely would have happened.

And you know, I never did find out how fast a Hellcat would go with water injection. Too busy watching my tail. Oh, well.

Not My Day

Told by William W. Wells
Captain, U.S.A.A.F.

After returning from combat duty in Europe in November 1944, I was very fortunately assigned as an instructor in P-47 aircraft and ordered to report to Dover Army Air Base in Delaware. In addition to the great assignment, I found the base commander was my commanding officer while I was in England, a really nice guy. Our job at Dover was to train the new incoming pilots in combat formation and tactics, low-level attack simulations, smoke deployment, dive-bombing, rocketry, and air and ground gunnery.

One morning, I was scheduled to lead a flight of four on a smoke screen laying out over Delaware Bay. We would usually pick a stationary target such as a dock, a lighthouse, buoys, and so forth, but this day I saw a medium-sized trawler, moving right along, and thought it would be good training to target it. The four of us spread out and dropped down to about 20 ft over the water. We came in to the rear of the ship and began to lay our smoke. The ship was moving downwind, so it proceeded forward in the smoke. We all congratulated ourselves on a job well done. We decided we would check back later and see how the ship was affected. In the meantime, it was time to take the boys up for a little high-speed, low-level simulated strafing attack. Our target this time was a group of huge chicken houses on a farm in Delaware. As you may know, Delaware produces lots of chickens.

We climbed up to about 10,000 ft. I told the boys to spread out, that we were going to build up speed, drop to the deck, and pretend to strafe the chicken houses, as if they were an enemy airfield. After reaching the altitude, we all did a wingover one after the other. We leveled off about 50 ft from the ground, speed near 400 mi/h. Shortly after we leveled off, I felt a sharp thump and noticed that part of my left wing was peeled back. We continued over the chicken houses, but I called the boys and told them we were going in to land. After landing, I found that I had hit a large goose and it had bored a large hole in the leading edge of my left wing. The chief of maintenance said it broke the main spar and he was surprised the wing didn't fold up on me. They had to put a new wing on that side.

That afternoon, the same day, I was scheduled to take up a four-ship flight on an aerial gunnery mission. It was a beautiful day with a few low-level, fluffy clouds here and there. Our mission was to start at 9000 ft and we were to peel off and fire our eight .50-caliber guns at a flat target being towed by another P-47. We were out over the Atlantic Ocean off the coast of Wildwood, New Jersey. We spent about 30 minutes, each taking turns firing at the flat target.

When we ran out of ammo, we regrouped and returned to Dover Army Air Base. We taxied up to the parking area, shut off our engines, and noticed a large group of officers approaching our aircraft. All of the high-ranking brass were there waiting for us. First, the Coast Guard had called and said the ship we had smoked had run aground. They were furious! Second, the owners of the chicken farm had called the base commander and said hundreds of chickens had smothered because of our buzz job, and they expected someone to pay. Third, and most serious, when we were firing aerial gunnery, the tow target pilot had erroneously drifted over the town of Wildwood, and some of our shots had gone through the roof of a beach house and into the bedroom, crashing into the bed where a woman lay. We really had some explaining to do here!

Of course, we didn't know this had happened because the clouds obscured the land area, and even the tow target pilot didn't know what had happened. Understand, he was the goat, but we took a lot of flak, too! This was truly not my day!

Five

Told by O. T. Ridley
First Lieutenant, U.S.A.A.F.

To every man upon this earth
Death cometh soon or late;
What better way to meet him
Than facing fearful odds
For the ashes of his fathers
And the temples of his gods.*

W e were in the vicinity of Madgeburg, Germany, 80 mi
southwest of Berlin, around 1200 hours at 29,000 ft.
The visibility was good. My P-51 was running well. I was flying wingman with Captain Leonard K. Carson, who at this
time had nine kills. He was an old hand, having been with
the 362d Squadron since its organization in the States.
Carson was well into his second tour with 88 missions.
He was about 5 ft 9 in tall with dark hair, brown eyes, and slight
of build. He was old, maybe 23. Carson was a farm boy, raised
in the vicinity of Clear Lake, Iowa. I did not know Carson well,
but well enough to know him as a loner. Carson was leading
blue flight, in the 362d's squadron of 16 P-51s. Our group, the
357th, was led by Major Joe Broadhead, the group operations
officer. The group was made up of three squadrons, the 364th,

---

*Lays of Ancient Rome—Macaulay, 1842—with slight modification. I found this quota-
tion in the book, *Combat Aviation*, by Keith Ayling, in the base library at Moore Field,
Mission, Texas, while I was in advanced pilot training. It has stood me in good stead.

363d, and the 362d, a total of about 50 P-51s. The group pilots had a habit of coming down from headquarters to fly with us when the mission was expected to be a productive one. Our squadron was led by Major John England.

The mission was briefed as a strafing operation against an oil depot in Annaberg, 45 mi southwest of Dresden, Germany. Our group was escorting another group, the 353d, also equipped with P-51s. If the mission went as briefed, both groups would strafe the oil depot. The 353d slowed their speed slightly and flew a B-17-like formation. Our group placed squadrons above, below, and beside the 353d group. We flew crisscross patterns above and around them providing protection as we normally did for the B-17s. The hope was that the German radar would confuse this large group of airplanes with a B-17 formation escorted by the usual fighter cover. Together, the two groups made a total of about 100 P-51s. The Germans took the bait! Our illustrious leader canceled the strafing mission, and we went for as many kills as fate would allow. The date was November 27, 1944.

Two large formations of enemy aircraft were reported. One of the large formations, still unidentified, made a turn and came toward us at the eight o'clock position. We released our drop tanks, which fell downward, and turned to meet the enemy. Our drop tanks were made of compressed paper, much like papier-mâché. The tanks held 108 gal of gasoline and enabled the P-51s to fly 7-hour missions. Normally the internal fuel tanks of the P-51 were used for takeoff and climb. Then we switched to the drop tanks and used that fuel going into the target area. One could not expect to do well in a dogfight with the tanks on because of the extra weight and the adverse aerodynamic effects. Carson continued to turn and dropped into the rear of a formation of over 50 Focke-Wulf-190s. The flights of our squadron were designated as red (leader), white, blue, and green, with four P-51s in each flight. Each flight flew in a fingertip formation. As we turned toward the German formation, the squadron separated into flights of four, then further into elements of two P-51s each. The element leader did the shooting; his wingman protected him and shot only if necessary or if invited. I dropped into an in-trail position slightly below and behind Carson. I was surprised to see no evasive action taken on the part of the Germans. We moved in closer to about 300 yd and directly behind the closest FW-190. There was a medium burst from Carson's guns, his tracers showed numerous strikes on the Focke-Wulf. The German started to burn and went into a turning dive to the left. We followed the FW-190 down to confirm the kill. The pilot was probably dead. The German crashed into the ground and exploded.

We climbed, near vertical, back up to the main formation of Germans, again closing to about 300 yd behind a Focke-Wulf at the rear of the formation, which was still intact. Carson fired two short bursts, getting strikes all over the fuselage. The Focke-Wulf started to smoke and then burn. The German dropped out of the formation and turned to the right until he was almost in an inverted position. He then went into a dive from which he did not recover. We followed him down to watch him crash and burn.

Once more we went back to the main German formation where Carson again took the nearest Focke-Wulf at the rear. This time he closed to about 100 yd and started to fire. With a few hits, the Focke-Wulf broke violently to the left and we broke with him. The Focke-Wulf did not turn tightly enough. This resulted in more hits on the FW-190's cockpit and engine. The Focke-Wulf started to smoke and burn badly. The pilot jettisoned his canopy and bailed out. He fell quite a distance. His chute did not open. We followed the FW-190 down and watched it crash and burn in the square of a small German village.

The area had now turned into the largest dogfight I had ever seen. It was a large spherical area of about 300 fighters. It ranged from the ground up to about 29,000 ft. As we climbed, again near vertical, back up through that ball of airplanes, Carson asked, "Ridley, are you still with me?" My answer, "Yes, sir." There were German pilots in their chutes floating through the melee. The dogfights were going on at all levels, vertical and horizontal, at every conceivable angle. On the way back up through the fight, I passed several Germans coming down by parachute fairly close. The German pilots wore a blue-gray uniform with black flying boots, and most still had on their helmets and black gloves. It was not customary to shoot Germans in their parachutes. However, there was a story rumored about the squadron that Major Pete Petersen of the 364th Squadron had witnessed an ME-109 pilot shooting a B-17 crew in their chutes as they left the burning B-17. The story goes that Petersen singled the ME-109 out, got on his tail, then pecked away at him with short gun burst until the German had to bail out. He then gave the German what he deserved and killed him in his chute. I came just close enough to these Germans to make them appreciate their health. When motivated, those Germans could climb the parachute shroud lines like a monkey.

Carson picked up the remains of the German formation about a mile ahead, and still balls out, we gained on the Germans. We then started down; Carson had seen a straggler. The FW-190 turned left and we gave chase for about 3 minutes before we came in range at 400 yd. The German turned sharply right, but again not tightly enough. Carson got hits on the fuselage. The pilot bailed out. As Carson was firing and getting hits on this FW-190, another formation of Focke-Wulfs, 40 to 50 airplanes, passed about 500 ft above and 400 yd in front of us. No attempt was made to come down and help the FW-190 being clobbered. They continued to climb out on a northeast heading.

Carson pulled up after the last engagement and set course for home. We had had a busy day at full power, and fuel was low. Then another Focke-Wulf made a run on us from a seven o'clock high position. We broke into him and he started a zooming climb. We gave chase and gained slowly at full power. Suddenly, he dropped his nose and headed for the deck. His course was parallel to railroad tracks below. As we gave chase, there was machine-gun and small arms fire from the railroad track. During the chase along the railroad track, there was the sound of a ping in my cockpit. My head was knocked from the headrest and I thought I could hear the rattling of metal in the cockpit. The engine continued to run well. All gauges were normal and Carson was about to get another one! He was now getting hits on the FW-190. The German started to smoke and then made a slight turn to the right. This resulted in more hits on the fuselage. The pilot jettisoned his canopy and we broke off the attack to the right. Carson waited for him to bail out, but he did not, so we turned back to engage him again. At about 700 yd, the pilot of the FW-190 pulled the nose up sharply and left the airplane. His chute opened.

Fuel was so low it was necessary for us to land in Belgium. Carson called a radio direction finding station (for DF steers)—I believe the call sign was "nuthouse"—and we were given a steer to a forward air base near Liege, Belgium, for refueling. The runway was a bit shorter than our 4500 ft at Leiston. As a member of the 362d Squadron, one always lands smartly by peeling up from the deck just above the landing runway, approaching a near loop, gear and flaps down, to complete the 180° turn and touchdown. Carson peeled up and to the left. I counted "one thousand, two thousand, three thousand" and peeled up good, tight, and steep. I was keeping pace with Carson. He touched down on the right. I was slightly to his left and a bit to his rear, a good position. However, as he slowed, I was a bit faster and passed him on my side of the runway. A faux pas. The runway ended at the top of a rise, and as I topped that rise, I found there a large pond of mud. I could see where many had gone before me. As I went into the pond, I added full power, held the tail down, and made a 180° turn out of the mud and back down the runway on my side of the runway. Fortunately, this maneuver had been explained to me in detail during primary flight training by my instructor, Mr. Ryan of Uvalde, Texas. As I passed Carson, going in the other direction—an even greater faux pas—he said, "Ridley, you are looking good." The mud was not splattered up on the wings and fuselage, apparently just on the underside of the P-51. After some taxiway maneuvering, I assumed my taxi position in trail and we were directed to our parking hardstands.

I shut the P-51 down, opened the canopy, unbuckled, and got out to check on the mud accumulation and possible damage. Captain Carson came over, and as we looked at the mud in the radiator and air scoop* and consulted with the maintenance

---

*The P-51 has a liquid-cooled engine. The air scoop and radiator are located under the fuselage and slightly behind the wing. The air cooler doors, within the air scoop, will control the flow of air through the radiator and thereby the engine temperature. Air cooler settings are manual and automatic.

man to learn the length of time required for cleaning, Carson said, "Tom, where did you get that tear in your parachute?" My answer: "Don't know." We went back to the cockpit of my P-51 to check it. There was a hole in the headrest just behind my head and there was a hole in the canopy railing, where a bullet of about .30 caliber had gone through. The bullet had apparently broken into three pieces; part went in the headrest, part into the parachute, and part fragmented in the floor area.

Back to the mud problem: It would be tomorrow before they could get the air scoop cleaned and checked. Carson's engine had been running rough since the last shoot down. This would also take time to diagnose and repair. We got to spend the night. Quite a few pilots had landed at the field as a result of the day's action. They put us in a block building, with cots close to the flight line, and served a dinner of good Army chow on a tin plate. There must have been about 20 in the same room. As a group we slept fitfully. Dinner did not agree with all of us. Some talked in their sleep: "break right green three," "five bogies two o'clock high," and from a distant corner came, "Ohoooo, Priscilla." Others snored or farted like Texas steers. In these strange surroundings, we stumbled over cots on the way to the latrine. Yet all in all, it was a good night. It sure beat being a guest of Adolf Hitler or even camping out.

The next morning, the weather was beautiful. Carson's P-51 had been repaired by changing 24 spark plugs and replacing the left magneto. My problem was another matter. The maintenance people said that the Belgian mud was particularly difficult. It might take several more hours to get the radiator clean. Carson departed for Leiston.* I was sorry I would not get to see him do his five victory rolls across the field at Leiston, one roll for each FW-190. The rolls are done, of course, after making a low pass across the field to bring out the audience. Then one comes back across the field at a very low altitude, in the grass, and proceeds to do his rolls.

I departed for Leiston about 3 hours after Carson's departure. With my P-51 refueled and the engine checking

---

*Leiston Air Base, Suffolk, England, was our home field, 20 mi northeast of Ipswich, England, and 3 mi inland from the English east coast.

okay, I taxied out and ran the air scoop manually through its full range and then checked it in automatic. The engine temperatures were okay, though I never knew and still don't know how they got all that mud out so quickly. I flew back to Leiston. It was a nice day across the Channel. The barrage balloons, with their cables, were up over Brussels. The English coast is always a pleasant sight on a sunny day. My landing pattern was just right and not too fast. I taxied back to the hardstand. The crew chief checked the headrest and canopy rail holes—they would be easy to repair. The parachute shop gave me a new chute and let me keep the pilot chute, with holes, from the damaged parachute as a souvenir. I still have it. The information services folks took my picture receiving a new parachute from parachute rigger Ernie Fliesh. They also did a short story about the engagement, all of which they sent to my hometown newspaper *The Fort Worth Star Telegram.*

Northern Italy, 1944

Told by George M. Blackburn
Major, U.S.A.A.F.

It was the first or second week of December 1944. The place was Grosetto, Italy. Lieutenant Lone Jones and I were sitting on a two holer outside the operations office prior to my 33d mission into northern Italy. It was cold. Lonesome was complaining about "sunny Italy." It snowed again the day before. We were both pilots of the 57th Fighter Group and the 66th Squadron.

The group had distinguished itself in North Africa with P-40s and fought and died brilliantly up the peninsula as General Mark Clark's Fifth Army had pushed the Germans northward. The group had most recently been on the island of Corsica and then to the cottages along the seashore. We were now housed in a small hotel inside the city of Goriest, which accommodated about 30 pilots.

It was early in the morning, about 0700, and we were uneasy. Lone Jones was to be the leader of this show of four aircraft. The mission was rather unusual, and I guess we were both a little edgy. As we sat there contemplating what was ahead, I felt like I was being potty trained. My feet wouldn't reach the floor but my flight suit plopped on the floor. It seemed that we should be treated better, hot shot pilots with accommodations no better than this. Then Lone Jones said that he believed his grandfather grew up and lived most of his life with an outdoor privy so we shouldn't complain.

It was then that I wanted to tell him about a dream I had the night before. I dreamed I had crash landed and hit my head pretty severely on the canopy which resulted in a huge knot on the left side. Lone Jones said, "Well, what else? Is that all there was to it?" I continued, "No. My guardian angel told me not to worry, that everything would be all right." Lone asked, "What's this guardian angel bit?" I said, "Well, I really believe in it. I think they are out there taking care of us. Ever since my English writing days at Vanderbilt University in Nashville, my old professor Dr. Duncan said I was a mystic and with mysticism seems to go the guardian angels." "Well, what's mysticism?" asked Lone. "Well, I really don't know. It's sort of a reverence for nature, for the birds and the animals, the spirit world, and along with it comes the guardian angel that you think will take care of you. Somehow I have always felt that I have been taken care of," I answered. Lone looked over at me and said, "Well, you're going to need a guardian angel on this mission. You had better put her in your cockpit cause this mission is going to get a little hairy. Let's get going to the rest of the briefing." With that we zipped up and headed for the operations office where Lieutenant Apostilou began the briefing about the upcoming mission.

"It's a first," he said, "and if we do it well, we will all be proud of the 66th. It's a tunnel mission and tunnel bombing. We have discovered, through intelligence, that the Germans are using an alternate to the Brenner Pass. They have a tunnel through the Alps and are running supplies at night. It is our mission to bomb this tunnel opening."

He went on to remind us that there were two things we would have to be careful about. The first was that we would be flying in at about 300 mi/h and we needed to remember the mushing characteristics of the P-47. The 7 tons of airplane would be determined to fly into the mountain at that speed. Second, we must release the bombs and immediately do a quick chandelle to the left to avoid our own bomb blast as well as to avoid colliding with the mountain. Those two things were emphasized, mushing and a quick chandelle out of harm's way. With that, Lone Jones said, "Let's do it," and we boarded the weapons carrier with our parachutes and were dropped off at our respective airplanes.

My crew chief, Sergeant Chip Burke, a good old Georgia boy, was waiting for me. "She's ready, Lieutenant. Hope to see you back in 2.5 hours." We taxied out, took off, and formed up in our four-plane formation. We headed north over the Apennines, the Po Valley, and onward to the foot of the Alps.

Our first problem was to find this opening in the mountains, which we knew was somewhere west of Lake Garda. Lone and I were circling the area trying to spot the elusive tunnel. Finally he spotted it. "It's camouflaged," he radioed. "They have tried to hide it but I see it. Line astern, go! Jackpot. Remember, watch your mushing." So as we prepared ourselves at the proper distance, Lone Jones peeled off and headed in toward the target. As he dropped his bombs, there was a tremendous cloud of dust and debris. I saw him chandelle up and away from the tunnel. His wingman soon followed and then it was my turn. I remember rolling over, leveling off, and releasing the bombs. Sure enough, the mountain loomed large in front of me. I immediately hit left rudder and hauled back on the stick. The P-47 went almost straight up. As we looked back, it appeared that we had been successful. The dust was climbing high into the sky and at no time did we see any flak. Lone ordered us to form up as we headed south toward home. We felt exalted because we had completed a difficult mission with no losses.

By this time, we were about 6000 ft and I advanced my throttle to ease into formation. Suddenly, my airplane took on a mind of its own. It flipped over on its left side and I found myself in a bewildering spiral toward mother earth. I fought and tried everything to pull out but to no avail. I was in a loose spin picking up speed. I heard Lone Jones call, "Blackie! What the hell is going on?" I replied, "I don't know. I am out of control." As the plane continued to spiral downward, I suddenly realized I had a serious problem. I didn't think

I could bail out because I was in a spin. Thoughts began to race through my mind just as I had read about when people thought they were going to die. I remembered my childhood, my mother and father, my brother and sister. I was ashamed because I hadn't been writing to my brother, Barton, who was flying P-38s on submarine patrol in Panama. (On January 6, 1946, one week before being discharged, he was killed in the crash of his plane.)

Even while these thoughts raced through my mind, I was not panicking. In fact, it was a somewhat calm feeling, but I knew it couldn't last much longer. I was beginning to see the tree lines. Suddenly, a voice or thought—who can distinguish?—entered my mind, "T-6 spin recovery, T-6 spin recovery." With that, I immediately chopped the throttle, hit hard right rudder, and popped the stick forward. There was an immediate response to the controls and I began to see a skyline instead of treetops. I was that low. As I began to ease up into the horizon, I realized that I still did not have control of the airplane. The left wing was dipping downward and I was flying somewhat in a circle.

About that time, Lone Jones pulled up and said, "Blackie, look at your left wing." I glanced out at the wing and there to my horror was an amazing sight. Both of the ammunition and gun box covers were flopped over against my left aileron. The air rushing over this open space was forcing the wing down, and I could not turn right because the aileron was blocked.

I found myself adding throttle to gain altitude, but as I gained more speed, the wing was being forced downward again. I was still in a slow circle to the left. I knew I had to gain some altitude because I would be a sitting duck for flak where I was. Lone Jones was circling above me and he said, "Keep trying to gain some altitude, and we will stay on top and try to get you past the flak." I gradually gained some altitude and realized I might be able to stomp on the right rudder, which put the airplane in a skid.

After flying that way for some time, I realized I had to cross the Apennine Mountains. I didn't think I had the strength to continue on the right rudder or the fuel to keep making circles around and around. It was also holding back the other three airplanes. As we crossed the Po River, Lone Jones called, "Blackie, are you going to bail out or are you going to try to take it over the mountains?" I answered, "No, I am not going to bail out. I have complete faith in this airplane. I don't think I should risk going over the mountains because I can't get enough altitude. I think I'll set her down somewhere below." "Okay. We'll circle. You're on your own."

At that time, I saw a canal that looked like it had concrete sides. It was not too wide. I decided that I would make my airplane a water sled. I figured that if I could just ease down into the water and let the wings slide along the edges, it would be a beautiful, smooth ride. So since I was turning left anyway, I banked around, looked it over, and decided that was where I would set her down. When Lone saw what I was up to, he was pretty upset. "You can't do that! Why are you trying that?" I answered, "Because I've got to go down. I am worn out and I think it will be a real smooth ride." So I banked around, cut the throttle, dropped my flaps, and eased down toward the water.

Suddenly, Lone Jones cracked, "Blackie, there's a footbridge ahead of you; it looks like two footbridges." "My God," I thought. I couldn't see them. I hadn't seen them during the approach, and now the big engine was in front of me and I surely couldn't see them. I tightened my harness, chopped the throttle, and eased her down. One last look at the airspeed indicated 110 mi/h. I decided it was time to hit. I pushed the stick forward and she smacked the ground. I had chickened out on putting it in the water for fear that it would slide too far and hit the bridge. I had elected to land on the grassy strip next to the canal.

When I came to a halt, I was a little dazed. I realized I had failed to open my canopy, which is a goofball trick. As I sat there, I realized that I had a pain coming from my head. On the left side of my head, I had a large knot as a result of having bumped the canopy. Of course, had I jettisoned the canopy, if I had rolled it back, it would not have happened. My dream, exactly as I had dreamed. Then I took the mechanical release, pulled the canopy back, and looked out at my left wing. At this time, it had broken loose from the fuselage and the ammunition was streaming behind it. Close examination showed an explosion of shrapnel had penetrated the leading edge of the wing. The forces had warped the cover plate thus popping the Dzus fasteners. The wind did the rest. A freak and lucky hit.

I flipped the switch back on to see if my radio would work and called Lone. "I'm okay," I told him, "When you get back, tell them where I am. Pinpoint it." "Well Blackie, we know where the Germans are but we don't know where the blackshirts are. That's your real problem, you know." We all had a young lieutenant on our minds because on his second mission he crash landed and the blackshirts had captured him. They dragged him through the town behind a truck, of course killing him. They were noted for their viciousness and they seemed to hate American pilots.

About that time, Lone called, "Blackie, there are five or six people running toward your airplane. There is one thing I do know, they are not Germans. I hope they are partisans. We will circle until we see how you are treated." As I looked to my right, sure enough these people were coming toward me talking in Italian. As they got closer, I shouted, "Are you partisans?" "Si, si," they replied, "partisana." What a relief. Suddenly, I felt like I was going to make it after all. I called Lone Jones one more time and told him they were partisans. He wished me well and I watched him disappear as the flight headed south over the mountains.

I suddenly felt very much alone. All those missions flying over this particular area; up there, we used to look down and wonder what people were doing and what was happening to them. Now I am down there with them. I am sure I will find out what they are doing. As I climbed out of the airplane, I decided I wanted my parachute silk and my clock. I pulled out a knife and began to cut the parachute seat off from the straps and then I used the edge of it to unscrew the clock and cut it out. By this time, I had stepped off the wing and the people were upon me.

There was a young girl about 25 or 26 years old leading them. She was speaking in broken English but she was obviously friendly. She had a white band around her head and there was a green star in the front. What that meant I had no idea, but I realized that they meant me no harm. One old lady came forward with a bottle. I presumed it was wine. She thrust it at me and made a motion for me to drink. I drank a good swallow and it about blew my head off. It was some kind of cognac and it did warm me up. The smiles and the talking and the gestures made me feel like I was a hero who had just come home. While all this was going on, I saw two of the women begin to point and run their hands over the canopy. They were excited about the canopy and I did not understand what it was all about. Then the girl said, "They want it for a washtub. Do you mind if they take it?" I said I did not mind. In fact, I hoped it would be a token of goodwill, perhaps make my stay with them a little nicer. Then the girl said, "We must go. We must go quickly. I will hide you at our house. We are happy to help you. We have watched you many, many times fly over and wondered who you are and what you look like. We know you are helping us defeat the Germans and you must know we are your friends." I felt real good about that.

We got to a small masonrylike house. All of the houses were similar. I never saw a wooden house of any description. We sat down at a table and her mother, I presume, offered me some black bread. It was as hard as a rock. Some more cognac, which I merely sipped. Then she began to talk. She asked my name and I said, "Georgio." "Oh," she said. "Georgio, good name. I often wonder what you Americans look like when you fly over us. Now I see. My name is Louisa. This is my brother, my mother, and my father. We have suffered from the Germans. They came through here, raped me, stole everything we had, took our one cow with them. Now we have no milk. It has been very hard as you can see; we have nothing. It is all Mussolini's fault. We hate him, too. I was once an educated person. I went 2 years to the University of Pisa. I have studied English since a little girl."

I told her I was mighty happy to know that because it made things easy. I had been very apprehensive when they approached me because we knew about the blackshirts. "Oh, yes," she said. "The blackshirts, they are bad, very bad. They are Fascisti; they will support Mussolini. Mussolini, no bono, no good. He brought suffering, misery, destruction to our country. And for what? Soon we will get him." I could see the hatred in her eyes as she said, "Soon there will be no more Mussolini." I believed her. Then she said, "Let me show you where you will sleep tonight." With that, we went outside. There was a small barn with some hay in the back. She showed me how to pull it away from the wall and said, "You must sleep here tonight. During the day, no Germans come; they hide. At night, they come out like cockroaches. They fill the road with supplies, delivering them to the soldiers in the mountains. They know during the day, if they dared, you would destroy them. We have seen it happen, so you must hide here at night and it will not be too bad." Then she motioned for me to sit down on a bench. I could tell she wanted to talk.

By then, I had a chance to really look at her. She was very pretty. She had northern Italian blonde hair and blue eyes. Her shoes were rough, with old socks around her ankles; her dress reminded me of the old Purina Startin' Chick Chow bags my mother used to make dishtowels out of during the early days of the Depression. Obviously, they had few clothes. She was still wearing the band around her head, so I asked her, "Louisa, what is this band and this star?" "It is the star of the partisani," she said. "We are well organized. We are everywhere. We kill Germans whenever we can. I have killed one myself." I was startled. How could she have killed a German soldier and still be alive? Then she held out her hand and said, "I killed him with scissors. He raped me. While he was raping me, I plunge these scissors into his back." "Lord," I thought, "this girl is brave, courageous, and determined. I am glad she is a friend and not an enemy." Then she said, "We will eat supper." She didn't use the word *supper*. I don't know what she used but anyway, "We will eat this night, then I take you to a place before you go to sleep I wish you to see." "What kind of place?" I asked. She said, "It is a trial." My heart jumped. A trial! Me? My trial? What have I done? What is she talking about? Suddenly, my sense of security evaporated.

She sensed my anxiety and grinned, "Not your trial. It is a trial of one of ours. A traitor. He delivered our friend to the enemy and he was executed," "What enemy, the Germans?" I asked. "No! The Fascisti," she said.

This seemed more like a dream. Being schooled in a system of justice with which I was familiar, I could not imagine attending such an event. As we rode in a horse-drawn cart, we suddenly stopped at a small church or maybe a chapel. We walked in and I stood out like a sore thumb. There were already 15 or 20 people there. They all turned to look at me. I could not tell from their expressions whether they approved or disapproved. I felt a little uneasy. She motioned for me to sit down.

There was a small stage up front with one chair and a door on the left and a door on the right. Suddenly, a young man of maybe 30 or 35 appeared from the left and then on the right two people appeared. One was being guided by the other toward the chair. He was extremely upset and distraught. His head was bowed and he was forced into the chair. I have never heard such a tirade of foreign expressions coming from the first fellow, obviously the prosecutor. He was shaking his finger, rolling his eyes, shaking his head. He was extremely upset, but so was the other individual. The fellow in the chair began to tremble; his hands were shaking and I saw sweat pouring down his face. Then the prosecutor turned to the crowd and obviously asked them for approval. They all nodded their heads and then they began to get loud. Many of them suddenly stood and shook their fists at the man in the chair. There was total disorder, and emotions were running at a fever pitch.

It was then that the young man broke into loud sobbing. Still shaking, he raised his head and with tears in his eyes he put out his hands and pleaded for his release, but to no avail. All he did was provoke more shouts and angry words. Suddenly, the young man was grabbed and taken out the door through which he had entered. A subdued hush fell over the group in the church. Some looked at one another, one started clapping, and it was picked up by the rest of them. They were all clapping until it was loud and had a very final tone about it. Then Louisa turned and said, "We will go." With that, we left, climbed onto the horse-drawn cart, and headed back to her place.

By then, it was dusk and she led me back to the little barn. She pointed to the hay and said, "It is time that you disappeared. I will see you in the morning and we will talk." Believing that her concerns for my safety were very real and that the Germans might well be coming down the road this night, I found a good place under the hay and stretched out to contemplate all that had happened to me on this day of my life. I congratulated myself because I knew it could have been much worse. I thanked my guardian angel for shepherding me through this very eventful day.

Contrary to what one might believe, I was able to sleep rather soundly that night. The next morning, I was awakened by a rooster and a sound and smell that was rather familiar from my childhood. It was the smell of ammonia from damp chicken droppings. Obviously, there were some laying hens around the barnyard. After the rooster finished his crowing, I raised up and stretched myself. Louisa was standing in the doorway. "We have breakfast," she said. "Come."

Her mother and father were very cordial and hospitable. They served fresh eggs, more black bread, and believe it or not, some wine. I had never thought about wine for breakfast, but perhaps it was all they had. We finished our breakfast and Louisa motioned for me to come outside and head back to the little barn. There we sat again on the bench and she said, "One of these days, we are going to kill Mussolini. We will for a very strong, young Communist party." It was then that I remembered seeing the hammer and sickle splashed on some of the masonry buildings. It had never occurred to me that Italy was going communist. How would I know? She obviously had some strong beliefs and was willing to sacrifice herself to pursue her goals.

Then she said, "They will be coming for you." I asked, "Who will be coming for me?" "Your friends," she answered. "They pinpointed where you went down. They are sending a small plane near where your plane is down and they will pick you up." I asked, "How do you know that? How in the world do you know that?" She said, "We have ways. I won't go into it, but we know." Then she seemed philosophical. She said, "You are the first American I have been close to, to talk to, and when you are gone I will always remember you. I will think about you when you are back in America and the war is over and we will be rebuilding our country again." With that, she startled me and for the first time she reached out and took both my hands and looked me straight in the eyes. "Georgio, I like you. I like Americanos and we thank you. All of us thank you for what you have done for our country." I don't know how I felt. I felt a certain amount of pride, of course, but I was also still a little startled by her philosophical conversation and by how mature and intelligent she seemed.

I don't remember how we spent the rest of the day but sometime around 1500 hours she informed me that they were coming in the evening and that I should be ready. She pointed to the parachute and the clock that I had laid down and said, "Don't forget these." Then she said, "There is a small pasture near your plane. He will land the little plane there and then take you back around the Apennines on the Adriatic side and back home to Grosetto." Then she shocked me. She said, "Let me see your pistol." I thought, "How does she know I have a .45 tucked in my jacket pocket?" She read me like a book. She said, "The weight is dangling on the left side of your jacket. I know you have it. I have heard about your pistol. I wish to see it."

I pulled out the .45 and showed it to her. "I wish we had many of these" she said. "It is very good." Then she politely handed it back to me. "I know you are with the 57th Group and by the little character on your plane you are the 66th Squadron and there are two other squadrons. You see, we know a lot about you. I also know that the Germans do not like you. They would never shoot you down in your parachute before, but now they will because you dropped that bad fiery jelly on them. They say it is horrible." I nodded and said, "Yes, it is called napalm and it is horrible. I don't know why some expert decided we needed to do that, but it was canceled very soon after it started because none of us felt that it was necessary. But you are right, we have been told by intelligence that we are in jeopardy if we bail out." As we talked, I could see that she was softening and it was almost like a girlfriend sending her boyfriend off to war. She changed from the fierce little fighter she pretended to be to a young woman almost weary of what had been happening to her all this time. I felt deeply indebted to her, but when I thought about her scissors story, I decided that was as far as I would like to go with this young lady.

About an hour later, I heard a small engine in the distance approaching from the east. Very soon, I recognized an L-5, the kind used for the Rover Joe spotting missions over the enemy lines so they could identify gun emplacements and direct our P-47s to destroy those emplacements. I was wondering who it would be. He circled and landed very close to the airplane, opened his door, and waved to me to come on, quickly. At that point, Louisa rushed up, threw her arms around me, and planted a kiss on my cheek. "Good-bye," she said. "Good-bye forever and good luck."

With that, I bounded for the little L-5 and climbed aboard. To my astonishment, there sat Lieutenant Richards, my cadet transition friend. We had been together the whole time. "Blackie, what are you doing with that girl? I thought you would be up here in bad shape. You been trying to put the make on that girl?" I turned to him and I said, "I'll tell you about her and her scissors a little later on. I guarantee you that I was trying nothing with that girl. She saved my life. I owe her." Back home that evening in the hotel bar, we drank a toast to the partisans of northern Italy because we all knew the many acts they had committed in our behalf. Of course, the boys were glad to see me and gave me a lot of teasing about Louisa.

The next morning, I climbed back into a P-47 and headed north again. On the way back, I dropped out of formation, found my airplane, circled, and buzzed Louisa. I saw her run out of the house and wave her arms. I rolled over and continued to roll until I had reached my altitude, wiggled my wings, and headed back to the field. That was my last contact with Louisa.

Sixty missions later, sometime in April, we had forced the Germans north across the Po Valley, and they were backed up against the foot of the Alps. We had moved our base of operations to a grass field, believe it or not, for that big P-47 in a little town called Villafranca, south of Verona. By then, we all knew the war was nearly over because there was very little resistance of any kind. One day, the word came back that Mussolini had been murdered. He had been caught on the road near Lake Como on the way to Switzerland and had been waylaid by partisans. They had taken $3 million from him and shot him, his mistress, and his bodyguard. History says that he died daring them to shoot him, and so they did.

Upon hearing this news, a friend of mine and I commandeered a little Fiat and decided to drive over to Milano to see Mussolini. Upon arrival in the town, there were hordes of people going by. I was shocked at what I saw. Being pilots, we seldom, if ever, saw the destruction firsthand that resulted from the bombing. I was suddenly face to face with three bodies. Bullet holes had riddled all of them, Mussolini in particular. They were tied and hung by their feet, side by side. It was not a pretty sight. I was suddenly struck by the prophecy that Louisa had made. She had said they would kill Mussolini, and they had. We were told later that the $3 million was seed money for the young Italian Communist party. On May 8, 1945, I was on a mission up the Brenner Pass north to Trento, south of Munich, when the radar people called and said not to drop any bombs. The Germans had surrendered. It was an exhilarating day.

Years later, when I bought my first son his first little animal, it was a female cat. I often watched her play and watched her movements. One day in particular, I saw her stalking a bird, and to my total amazement, as the bird flew, she leaped into the air and caught the bird with her extended claws. She pulled it to her mouth and looked up at me with those blue eyes. She then dropped the bird and started rubbing against my legs purring.

We had named her Louisa.

My Squadron's Pacific Flight from Eniwetok Atoll to Okinawa

Told by Hensley Williams
Major, U.S. Marine Corps

This story is not about combat but about my squadron's (VMF-113) flight over water from the Pacific atoll of Eniwetok (actually Engibe Island) to Okinawa.

In late 1944 and early 1945, we started making preparations for our movement from Eniwetok to Okinawa, a flight of about 2500 kn, all over water. First, it was over 1000 kn to Saipan, then over 650 kn to Iwo Jima, and finally about 750 kn to Okinawa. My squadron had 32 F4U Corsair fighter planes with which to make the trip. We also had four R5C transports to accompany us for navigational and air-sea rescue purposes. Eight Corsairs were to fly wing positions (four on the right and four on the left) on each of the four R5C transports. We left Eniwetok for Saipan on a beautiful day in the early spring of 1945 (all 36 airplanes). All went well for about 200 kn when the transport plane flight paths began to spread so far apart they were barely visible. I called the pilots of the three transports (other than the one I was flying wing on) and told them to move in closer to us. They replied their navigation was correct and they were going to continue as they were now doing. I called them again and told them I was in command of the entire flight and they had better follow orders. There was no response, and soon the other three transport planes along with their 24 F4Us were completely out of sight. I called a third time telling them this was the last call, and if they did not obey, I was recommending them for a general court-martial as soon as we landed. Again, I got no radio response, but soon they all complied and were back in their assigned positions!

The flight from Saipan to Iwo Jima was unusual in that twice we got almost to Iwo Jima and had to turn back because of a very heavy weather front. On the third try, we again hit a front but managed to pass through cloud layers at 14,500 ft, no small trick for a formation of 32 fighters and four transports.

The flight from Iwo Jima to Okinawa was delayed for 3 days due to the weather. On the fourth day, the weather was fine at first, but as we approached Okinawa, the ceiling got lower and lower until we were down to 100 ft and visibility was poor. Therefore, we had to break the squadron up into eight divisions of four fighters each. It was truly remarkable that all 32 fighters landed safely on three different airfields! Detailed advance planning and information on the area really paid off. I will never forget this experience.

Incidents
Remembered
50 Years Later

Told by Johnnie B. Corbitt
Captain, U.S.A.A.F.

On April 30, 1944, two hundred former Air Corps Flying School instructors boarded a car ferry in New York harbor for a ship awaiting us further out in the harbor. On the other side of the ferry were an equal number of nurses. We thought, "What a great war!" The ferry headed for a ship and the nurses began to unload. We grabbed our luggage and were ready to unload also. The Army transportation sergeant smiled and said, "Sorry, boys, you don't get off here." The ferry proceeded to another ship already loaded with 5000 infantrymen. What a letdown.

Our ship was the S.S. *Sterling Castle,* a Cunard White Star Line luxury liner whose prewar passenger route had been from Southampton to South Africa. The crew still wore its company's uniforms and the waiters in the dining room had tea towels on their arms. The tables had linen tablecloths and the ship's best service. The food was British-style fish, served twice a day. Occasionally, we got some white bean soup! I ate a lot of Hershey's chocolate purchased from the ship's canteen.

We were in a very large convoy and it took 14 days to cross the Atlantic. Our ship used only one of its two large engines to keep up with the convoy. We had some excitement one day when several Navy destroyers raced around dropping depth charges at reported U-boats. We never saw any, but they got our attention. Near Ireland, they started the other engine, and in an hour, we had lost sight of the rest of the convoy and were soon in Liverpool. We were sent to Stone and Atcham for processing and on to Shrewsbury for 3 weeks of orientation flying.

We boarded a train, headed south to the town of Isbley, and reported to the 48th Fighter Group of the Ninth Air Force on June 5, 1944. Sleep was difficult that night as there was the constant roar of aircraft engines all night. I thought, "They really fight the war down here in southern England!" When we got up the next morning, June 6, 1944, the sky was covered from horizon to horizon with planes and gliders. A driver came to pick us up and yelled, "They invaded the Continent last night." It had finally begun.

After breakfast, we reported to the 493d Fighter Squadron. They were so excited and busy that they hardly knew we were there for the next several days. I was given two orientation flights on June 10 and still waited to see action.

I flew my first mission on June 15 late in the afternoon and it was nearing dusk, making the tracer rounds that they were firing very visible. I thought, "They are firing real bullets at me. They are trying to kill me. I could get hurt. This is not fun and games anymore." The reality of war had set in. Missions came with great regularity from then on.

One mission I remember turned out to be a nonmission. When I went to get in the plane, I noticed that the armament crew had failed to load bombs on my plane. I said nothing because I wanted to go on the mission and my machine guns were loaded. Major Bryson, the squadron commander, noticed the missing bombs after we got in the air and quite brusquely sent me back to base. That was the only mission that I failed to complete during the war.

During the winter of 1944–1945, the buzz bombs, or pulse jet flying bombs, were a source of constant annoyance. As long as you could hear the putt-putt of the engine, everything was all right. When you heard the engine cut off, you knew it was going to land and explode in a matter of seconds. Our base at St. Trond, Belgium, was right under the alert flyway to Antwerp, which they were trying to knock out as a main supply source. Once, as I was sitting in my plane on runway alert, I glanced up just as a buzz bomb was a few feet off the ground. It hit immediately on a spot about halfway between our gasoline dump and the ammunition dump.

One mission, while we were stationed at Villacoublay Airfield in Paris, went deep into Germany to cut rail lines and destroy rolling stock. We carried a full load of fuel with two external fuel tanks and a 500-lb bomb. It was about a 4-hour flight, so by the time we got back to Villacoublay, we were all nearing the empty mark on all tanks. The atmosphere was very hazy, making landing difficult. Looking straight down was okay, but you could not see the runway as you turned on final approach. A 16-plane squadron was trying to land ahead of us and was having trouble. We circled and circled and got lower on fuel. The tower officer yelled at a plane on final approach that his gear was not down and to go around. He landed, wheels up, in the middle of the only usable runway. Someone muttered in his microphone, "We ought to strafe the SOB." It was finally cleared and all planes got down safely, but some ran out of fuel while taxiing back to the squadron areas. There was a blacktop highway about half a mile west of the runway, and in the thick haze, several planes tried to land on the highway, but eventually realized what it was and managed to avoid a bad situation.

The Stars and Stripes reported that during the raising of the flag over the Rhine at Cologne, eight P-47s circled overhead to protect against a sneak attack on the ceremonies. I had the privilege of leading that mission.

We flew top cover for the airborne crossing of the Rhine north of Cologne. We encountered no opposition in the air, but the antiaircraft fire from the ground was fierce, especially that directed at the troop transports and gliders. There was a feeling of helplessness and horror on seeing large numbers of planes and gliders loaded with troopers being shot down, crashing, and burning on the ground.

There was one aggravating mission I remember very well. We were preparing to take off and a broken wire from the wire matting on our runway punctured one of my tires. My wingman had no training or experience in leading missions, so we swapped planes and I continued on with the mission. I navigated by ground checkpoints on the way to the target. With business finished, I set a compass heading back to our base and relaxed, knowing that I could recognize our crossing of the Rhine. When I got to the Rhine, there was nothing recognizable in any direction. I finally headed west and eventually saw an airfield with lots of P-47s in the parking area. I led my flight to a landing and learned that we were at a field near Metz, France. They checked the compass and found that it was many degrees off its correct setting. We spent the night and went back to St. Trond the next morning. I found out that the wingman knew it was not correct but he didn't do any navigating.

After our ground forces had captured and crossed the bridge at Remagen, our job was to protect the bridge and help clear the way for the ground forces' sweep across Germany. Our forces had fanned out for about 10 to 15 mi after crossing the Rhine, and we were pounding the German's front line out in front of our troops. As we were coming off a strafing run, someone in the area yelled, "There's a jet in the area." I had never seen one, but directly in front of me, about 5000 ft above, was unmistakably a jet airplane. No propeller. Weird. I told the flight to follow me, I gave it full throttle, and tried to close on him. He was loaded with bombs and headed for the Remagen Bridge. He saw us, jettisoned his bombs, and turned south toward the German-held city of Koblenz. Even cutting across toward him, I could not gain on him and he was leading us toward a pass over Koblenz. He was obviously baiting us, so I broke off pursuit and headed back to base. I always regretted not firing a burst at him. Even though I was out of range, I would have gotten him on the gun camera. We possibly saved the bridge for another day or two of use in getting our men across the Rhine River.

My friend Ed Miller tells a story of when he was the air-craft coordinator in the lead tank of an armored column we were working with. The tanks were being held up by a German tank hiding in a village and zeroing in on the column. As per procedure, he called for assistance from fighter aircraft in the area. I responded a fixup (our call sign) red leader, and he recognized our call sign because he was a member of our squadron when he was on flying duty. He gave us the coordinates of the tank's position. We immediately located the tank, went in, and took it out. There was great rejoicing on the ground and Miller called and asked us, as a morale booster, to come back over and give them a show. The flight leader came back on a low-level buzz job and he pulled up and away and did a roll as a salute to the men. The men whooped and hollered and Ed said, "That was my buddy, Johnnie Corbitt." I remembered the mission and everything except doing the roll. A common occurrence is for one person to remember an incident so vividly, but another person involved will not remember a thing about it. I guess that 50 years make it excusable.

Often, when we had a few hours off, we would go to Hasselt, Belgium, and get a chopped steak (probably horse meat, but good) and a large platter of French fries. The area was a rest area for ground troops back from the front. One time, we went into a large hall with a bar across one end that was jammed with troops still in battle gear. They were a right rough looking bunch. They looked suspiciously at this bunch of kids in Air Corps uniforms, and one finally asked what outfit we were with. When we replied, "Fixup Squadron," they rushed us and said, "You guys work with us all the time." For the rest of the evening, we could not buy anything. We were treated like royalty. It was nice to know that the people who really counted appreciated what we were doing. One comment from them was, "Boy, I wouldn't be up there, where you guys are, for anything." I was thinking the same thing about being down there, where they were.

January 1, 1945. We had a big New Year's Eve party and we were not expecting to fly the next day because of the weather. But early on the first, we were routed out for a mission. I was sitting in the cockpit of my plane, chatting with my crew chief, Sergeant Kendrick. He asked, "What kind of plane is that on the horizon?" "That's a Spitfire. No, a Typhoon. It's a 109!"

He jumped off the wing and headed for the foxhole. I unbuckled the safety belt and parachute, jumped out, and headed there myself. The foxhole was L-shaped, and being the last in, I was in the angle of the L. Whichever way I stuck my head, I felt that my rear end was exposed to the direct line of fire. We were fortunate that none of our planes were severely damaged, and we were able to mount a full-scale mission immediately. Some other squadrons were not able to send a full squadron of planes. One German plane was shot down and the pilot bailed out and landed very near squadron headquarters. Some of the enlisted men rushed out and captured this very young pilot. They marched him back to our headquarters and I think some of the men were ready to carry out a sentence on the spot, but cooler heads prevailed.

The following items I just mention briefly, without detail:

1. The character who broke up formations by yelling "flak" sounded like the same person throughout the war.

2. The helplessness on hearing a pilot scream, "They've shot off my leg."

3. Seeing the parachute from an exploding B-17 drift back into German territory.

4. A 5-day pass to Brussels just days after the liberation in September.

5. A 7-day leave to the Riviera in February 1945.

6. J. T. Whitlock borrowing from and repaying money to me to play poker.

7. Saving all of my mission whiskey and giving it to my crew when I got a quart.

8. Bryson and Baker spotting bogeys and identifying them before the rest of us could spot them.

9. Captain Harrell sleeping in full uniform, including steel helmet, during the Battle of the Bulge.

10. Newburn arguing with Stone, Franks, and Underwood over the merits of Texas and with anyone who would argue the merits of the South.

11. First order of business after a move: "build a club."

12. Two shot down on a mission against trapped Germans near the Ruhr River.

13. George Jones's belly landing in no man's land in front of me and J. T. Whitlock bailing out with shrapnel through his elbow behind me.

14. Brush Hall's hat routine: last thing off at night, first thing on in the morning.

15. Howard Foulkes's photography shop under the blankets.

16. Dear hunting around Illesheim near the end.

17. Heading for dugouts on A-4 when the shrapnel started falling.

18. B-17s landing at St. Trond, some okay, many crashing and burning.

19. Eighth Air Force fighter landing and asking, "Where's the club?"

20. The contrails of hundreds of "heavies" en route to Germany and the smoke trails of the damaged coming back.

21. First night in Paris at the Sphinx Club.

22. The bus-truck trip from Illesheim to Brussels, following the Rhine to Cologne. I don't remember the trip back.

23. The smell of death in the trenches at Normandy.

24. The hot chocolate and the deep-fried cheese sandwiches in the ready room.

25. The day the pinochle deck got into the blackjack game.

26. The women attendants in the men's rest rooms in Paris.

27. Short of cash? Partial payments in Paris and Brussels.

28. Leaflet missions. Oh! How we hated them.

29. Ferrying our beloved P-47s to Paris for disposal at the end.

Not My Day II

Told by William W. Wells
Captain, U.S.A.A.F.

At Dover Army Air Base, we were busily training new pilots, getting them ready for duty in the Pacific. Aerial gunnery was probably one of the most important aspects of our training. Since the recent fiasco at Wildwood, New Jersey, in which some of our shots went through the roof of a beach house, we had been ordered to move further out to sea when firing gunnery for obvious reasons. About 3 weeks after the Wildwood incident, I was told to take a four-ship flight up for aerial gunnery about 30 mi off the coast of Cape May, New Jersey. It was a beautiful sunny day with a fairly strong wind off the ocean and not a cloud in the sky.

The tow target ship led the way out to the range and climbed up to about 7000 ft, and we were to go up to 10,000 where we would begin our passes. This time, we were to fire at a vertical target and, after building up our speed as we dropped down from up high, we approached the target from the side. Each of us had different colored ammunition so we could tell who was hitting the target. We took turns firing, and after about 20 minutes of this, one of my wingmen called and told me bluish-gray smoke was coming from my supercharger exhaust. I quickly checked my oil pressure gauge and noticed it had begun to drop. I immediately headed for shore, for by now we were probably about 50 mi out to sea.

When I headed for shore, I was at 9000 ft. I told my wing-man to call the Coast Guard air-sea rescue on the emergency channel because I fully expected to bail out. Gradually, I was losing power and the engine was running rough. My airspeed was dropping and the oil pressure was getting close to zero. By now, my airspeed was about 170 and my altitude was about 5000 ft. The plane was vibrating and shaking so vio-lently that I thought the gun sight was going to snap off. When I reached 3500 ft, I called my wingman and told him I was down to 140 mi/h and when I got to 2500 ft I was going over the side.

By now, I was about 10 mi offshore, so at 3000 ft, I popped my canopy, unhooked my radio and oxygen, undid my safety belts, lowered my seat, and set the trim so the airplane would head out to sea. By now, the engine was locked up and the prop barely windmilling. At 2500 ft, I crawled over the right side and fell out headfirst. The windstream was still over 100 mi/h, so it tore off my helmet. The wind grabbed my seat pack parachute and shifted it way up on my back, so when I tried to grab the rip cord, it wasn't in its regular position. I fell about 1500 ft before I found the rip cord, so I estimate the chute opened at about 800 ft! Then an eerie silence surrounded me. All of a sudden, the plane turned back toward the land and headed for me! It swished by me and missed by about 30 ft. I thought I was a goner! It then spiraled on down and crashed on the beach in flames. In the meantime, the 25-kn wind had blown me toward the beach and I feared it would blow me into the flames of the crash! I landed on the sand about 100 ft from the fire. I gathered up my chute and headed for the beach road. As I looked about me, I saw two Coast Guard cutters offshore, a dirigible from Lakehurst Naval Air Station, and a big twin-rotor helicopter above me.

When I reached the coast road, I saw a Jeep coming toward me. It stopped and the woman driver offered me a ride. I got in as she said she was headed for the fire and, at that point, she didn't know what had happened. About this time, the fire engines came to the scene. I thanked the lady, and the firemen said they would take me with them. They told me the lady in the Jeep was Pearl Buck, who had a beach house nearby. Finally, the big helicopter lowered a sling seat, pulled me up into it as it hovered, and whisked me back to base. Once again, this was not my day or maybe it was, considering what could have happened.

Pandemonium

Told by William S. Miller
Captain, U.S.A.A.F.

Lieutenant Colonel Fred Hook was leading red flight, six P-47s, after having completed Bill's five-mission combat training program. They were flying out of Ankang in central China. The mission today was to bomb and strafe the big airport at Hankow. They were armed with a 500-lb bomb under each wing, their usual load of .50-caliber machine-gun bullets, and the ever-present 75-gal belly tank filled with gasoline to stretch the range. Flying time to Hankow was only 1 hour and 45 minutes so they could have made it without the extra tanks, but they were hoping to run into some airborne Zeros, in which case the extra 75 gal would come in very handy. At full combat power, the big Pratt and Whitney R-2800 engine could suck up fuel at nearly 400 gal/h.

Fred—they used first names rather than ranks to confuse the Japanese as to where the higher ranking troops were—had an import guy from the group's 93d Squadron flying his wing. Fred was the deputy group commander of the 81st Fighter Group. Leading his second element was Bill of the 91st Squadron with Flight Officer George Frieze from the same squadron on his wing. The other two ships included Lieutenant Roy Holton of the 91st and another guy from the 92d. The trip to the enemy lines was uneventful except they all observed La Hokow Airfield as they passed it, and they commented that it would be the first recovery or emergency landing field on the way back from Hankow.

As they neared Hankow, they climbed up to 10,000 ft and checked carefully and hopefully for enemy airplanes. If they spotted them, they would jettison their bombs and belly tanks to streamline their planes and have a go at the enemy. When they flew over the airport, they were met by occasional bursts of 90-mm antiaircraft fire, but this didn't worry them too much. High-altitude antiaircraft guns were ineffective at best, and they were well below their best operating altitude. The only antiaircraft fire that really bothered them was the 20- and 40-mm cannons with an effective range of 1500 to 7000 ft and the .30- and .50-caliber machine guns, effective from 100 to 5000 ft. None of these would bother them until they started down on their bombing runs. They would be fairly vulnerable since dive-bombing required the pilot to fly a straight and predictable path in order to hit the target.

It was with great sorrow that they discovered no airborne Zeros. They then turned their attention to the ground. They saw no planes on the ground either, but they were pretty high to spot them. Fred told the flight to go into a loose string (follow the leader) formation and indicated that his flight would hit the northern hangar. He directed Bill's element to hit the operations building in the center and the other element to strike the southern hangar. He cautioned them to switch over from belly tanks, if they were using them for fuel, and told them to be sure and arm their bombs. This arming caused a rig to grab and pull a wire from the bomb's fuse as it fell from the bomb shackles and allowed the small propeller on the nose to start spinning. When the bomb fell approximately 100 ft, it was armed and ready to do its job when it hit the ground.

On the way down, they spaced themselves and rolled into an 80° dive. Next they started drifting their gun sights onto the appropriate targets and reduced power slowly on the way down. This was necessary to hold the speed down to a reasonable 380 mi/h. Otherwise, they would have built up to 450 to 460 mi/h and the air would have gotten too rough for good bombing accuracy. As Bill passed through 5000 ft, he sensed that the light cannon fire was intensifying, and at 4000 ft, he began to feel the pitter-patter of small bullets hitting his plane. A hit by machine-gun bullets feels and sounds like small scattered sleet and hail pellets striking an automobile. At 3000 ft, Bill lined his sights carefully on to the center of his target and continued to fine-tune his dive until he passed 1800. Then he toggled the thumb button on top of the control stick to send the bombs on their way. After this, he sucked the control stick back into his lap to maintain a 5-G pullout. This got him almost out of his dive at 800 ft, at which time he opened the throttle to full power and slashed rapidly down toward the ground. Next he dropped a wing to look for damage at the targets.

He counted 11 hits and was wondering what happened to the other when George said over the radio, "Bill, I think I have a hung bomb." Bill replied, "George, when we get out of groundfire range, I'll check you over." Sure enough, when Bill was able to do an S-turn to let George catch up, he found that the big 500-lb bomb on George's wing was flapping in the breeze and it was hanging on by only the rear shackle. Fortunately, the arming wire was still in the propeller on the nose of the bomb or George would not have been there. If the wire were to pull out in the slipstream, it would be good-bye George and anyone within 600 ft of him. Bill confirmed to George and the rest of red flight that he did have a hung bomb but did not mention the flapping arming wire.

After discussing the situation, Fred and Bill decided to continue on their planned strafing mission. They advised George to try to shake the bomb off by violently yo-yoing the stick. They told him to stay above 500 ft on his strafing passes and cautioned the other flight members to carefully avoid flying directly behind him just in case the bomb did let go. Today was not George's day! The bomb refused to sling off from the violent maneuvers, and when he started in on a 30° dive, a Japanese gunner got lucky. George experienced a sudden power loss, his big radial engine started making horrible grinding noises, and he started getting black oil all over his windscreen. Oil was spewing out of his top rear cylinder, and very soon he was not able to see outside of the bird. Bill told

George to go on instruments immediately, to take up a 290° heading toward home, and advised him to hunt around for the lowest power setting that would keep him airborne. Bill advised that he was going to head directly for La Hokow with George or at least get him as far away from Hankow as possible.

Everyone agreed that it was imperative to get George as far away from Hankow as possible. Japanese soldiers, or anyone's, tend to be downright hostile toward downed pilots who have just bombed or strafed them. The case of Roggy Rogenbauer was close on all of our minds. Roggy had been shot down over Hankow last month, and the Japanese had dragged his body around the city by a rope tied around his neck behind their version of a Jeep.

As they struggled away, Bill had George jettison his belly tank to reduce drag and told him to experiment with various power settings to reduce the load on the engine and to smooth it out. George found that it ran best at 1800 revs and 20 in of manifold pressure. Even so, the engine would occasionally shudder, misfire a few times, and shake the whole plane. It reminded Bill of a dog shaking a rat each time it happened. This power setting would normally have given George 170 mi/h and would have delivered 1200 hp. With the dead and dragging cylinder, it was only producing 900. With the extra drag of the hung bomb and the sick engine, he was only able to get 140 mi/h. At this speed, the plane was not its normal, stable platform, and George was having a helluva time flying instruments. Bill decided that George would never make it all the way home on instruments, so he told him to crack his canopy to see him out the left side. He got a little oil in his face this way, but it was far preferable to flying instruments in a wallowing airplane. That was not the only problem. In order to fly formation, the wingman needs to constantly change power settings. Every time George moved the throttle, his engine threatened to roll over and die.

Bill thought about the problem and improvised a solution. Normally, the wingman flew slightly behind his leader and made the necessary power adjustments to maintain that position. What if they reversed that procedure? They experimented with that for a while and found that it was no trouble for Bill to fly a formation position ahead and to the left of George. It felt weird but it worked fine and it let Bill make the critical power adjustments.

They knew the P-47 held 27 gal of oil and Bill knew from personal experience that the engine would hold together for at least 30 minutes even after the oil pressure gauge registered zero. However, the difference between his and George's case was that his engine had been running smoothly whereas George had internal damage. George's cylinder head temperature indicated that the engine was too hot, his oil pressure was fluctuating badly, and he could hear the clatter of metal on metal. At 1800 r/min, each of his 18 cylinders was compressing and firing 900 times per minute, or in the case of his top cylinder, it was trying to fire. Each time the piston came up, it resulted in squirting an ounce or so of oil onto George's canopy. Only God knew how much longer it would continue to run.

Another problem in the form of the first range of hills was looming on the horizon. Somehow, they were going to have to climb to at least 5000 ft to get over this group of mountains to get into La Hokow. Bill told George to hold his revolutions per minute at 1800 but to see if he could add 3 more inches of manifold pressure. By experimentation, he found that he could get way up to 150 mi/h or that he could hold 140 and then climb 100 to 120 ft/min. This might let them do it. Whereas before, as he pulled the nose up to try to climb, it only resulted in the plane slowing down, he could now actually climb a little. By now, they had been en route for 15 minutes and were about 35 mi from Hankow. They had about 100 mi to go to La Hokow. If they could only climb another 4500 ft to clear the mountains or even if they could get up another 500 ft, they had a chance of George living through this, Bill mused. But they were barely scraping along right off the ground. He was actually lifting up on his stick as if that would pull them up a few more feet. He didn't tell George, though, for there was no use worrying him while he was fighting to live.

It took 10 minutes of steady fighting for them to struggle up to 2000 ft. Unfortunately, the mountains were looming up toward them at a faster rate than they were able to climb. Bill debated about doing some side circling climbing turns but rejected that solution because even the shallowest turn would reduce their rate of climb to nothing. He could also have flown on a parallel course to the mountain range, but that would have resulted in taxing the engine further and still not getting them any closer to friendly territory. It was a helluva time to go hedgehopping, but if they could find the river a few miles ahead and if Bill's guess as to their position was correct, then it would lead them into La Hokow. Hedgehopping or buzzing at low level was one of their favorite sports, and they often did it on the way home from a mission if they did not have battle damage. The hedgehopping today was not any fun at 140 to 150 mi/h, and in the unwieldy reverse formation which they were flying, the big P-47 turned like a lead sled. When they came to a cul-de-sac during normal buzzing, they could pour the power to the bird and zoom out. But mushing and lumbering as they were doing now would allow them to zoom maybe 200 ft at the most. It was going to be close.

George must have sensed Bill's debate with himself and the doubts that he had as to their survival, for he called, "Red 3, how are we doing?" Bill knew that it was urgent to keep George with a positive attitude, so he replied, "We are doing fine, George." As he stretched the truth, they were 50 ft above the tops of the trees. Bill considered telling George to boost the power up a little bit but decided to let it alone. More pressure on the engine could cause it to quit at any time. Occasionally, Bill would lapse into a "what if" mode. For instance, what if that flapping arming wire came out of the bomb's propeller? They would both be blown to smithereens. "Oh well, you can't win them all," he thought.

By the time they came to the friendly river, they had struggled up to 3000 ft. With a sigh of relief, they headed northwest, up this river. It was the Han, which would lead them right past La Hokow and on to Ankang if La Hokow was weathered in or they couldn't land for some reason. The Han was a major tributary of the Yangtze and it was about the size of the Tennessee River in Bill's home territory. Most of the turns were large sweeping ones and there were no waterfalls or steep rapids. After a few minutes, they found that the mountain range climbed at almost the same rate that George's sick bird was climbing. Every time they gained 100 ft, they covered about 2.5 mi. This was going to let them make it if the engine held out and the bomb didn't blow.

They were doing a lot more talking today than they normally would have. Bill was doing this to keep George's and maybe his own mind off their problems. Soon they were up to at least 600 ft above the old Han River, and Bill was pleased to tell George that at last they were high enough for him to bail out if the engine quit. George said, "Hell, I hadn't even thought of that, until now."

When their first hour from Hankow rolled around, they were about 150 mi away, and blessing on top of blessing, they were past the bomb line, back in friendly territory. They had only 50 mi to La Hokow and it was now time to start figuring how to get George down safely. The foremost consideration was that there was about a 70-30 probability that the bomb would fall when they touched down. When it fell, there was a 50-50 probability that it would blow. Another major factor was whether the engine would continue to hang together. It was still spouting oil on each stroke, so oil starvation was still a few minutes off. Another bother was that neither of them had even thought about doing a formation landing in this goofy reverse formation. Yet another problem was that Bill had to make his formation power-on approach in such a manner that George did not have to change his power setting because they both knew that if the engine even sputtered near to the ground, he was a goner. There was also the nagging awareness that the approach and landing would have to be perfect the first time because a go-around or an aborted landing was impossible.

They discussed all these problems, and Bill said, "George, you better bail out. You have a much higher chance of living through this mess that way." Not only would George have a better chance, but Bill would too. If the huge bomb fell off, and if it blew, they were going to roll right into the bomb blast and not even a P-47 would survive that. George thought for a while and finally replied, "I agree it would be safer to bail out but, hell, it's my life, so if you don't mind, I'd rather take my chances in this airplane. Okay?" Bill replied, "You got it George. We'll give it a try."

La Hokow was coming up on the horizon, so Bill explained their problem and requested landing on the north-south grass runway. The tower informed them that runway was under construction and was not usable. They recommended using the main east-west runway and gave their approval for a landing to the east. Bill explained that there was almost a certainty that the huge bomb would fall off on landing and that it might very well blow up. He knew the east-west runway ran parallel to the ramp that contained the other airplanes, and it was also very near to the hangars and the personnel areas. Only God knew where the bomb might go when it came off, and it might blow some of those vulnerable people and things up with it. Another important factor was that the grass would be much softer and it would cushion their fall and that of the bomb, so Bill exercised the pilot's option and said, "La Hokow Tower, I intend to land on the grass to the west of the old north-south runway. We would appreciate it if you would clear the pattern when we are on final approach because we can't go around." The tower finally saw the wisdom of this solution and promised to cooperate.

The two aircraft started a slow descent and a gentle maneuver into an approach position to the airfield. George was able to reduce his power to 18 in of manifold and this gave him almost a perfect power setting to make his approach for landing. Many miles further out than they would normally have done it, they dropped their landing gear and this lowered their speed to 120 mi/h with an 800 ft/min rate of descent. Bill maintained a constant stream of patter to keep George good and loose during their descent. He told him he would give him a head nod to cut power completely when they were about 3 in above the ground. They would touch down wheels first in more or less of a level flying position rather than using their normal three-point landing. George was also told to brake his plane as hard as possible as soon as he touched the ground and to follow Bill's veer to the left as hard as possible to get out of the bomb's path, if or when it came off.

At 150 ft, when they had the airfield made for certain, Bill called for half flaps. This allowed them to raise their noses to nearly level flight positions, bleed off airspeed to 90 mi/h, and reduce their sink rate to 150 ft/min. When they got within 20 ft of the ground, ground effect started taking over, and they were able to raise their noses to 5° above the horizon. With this change in pitch attitude, their airspeed bled off to 75 mi/h and their sink rate lessened to 100 ft/min. At 3 in off the ground, Bill gave the signal for George to chip the power and told him to hold tight. They touched down in a nearly perfect wheel landing and started braking hard. The blessed big old Pratt and Whitney engine had come through one more time. They raised their flaps on a nod from Bill's head, and as they started to slow perceptibly, the bomb fell off and went skittering across the Tarmac. They swerved to the left, out of harm's way, as hard as possible without shearing off their landing gear.

Fortunately, the main landing gear on the P-47 was built like a tank and it hung together. They both breathed a prayer as they passed to the side of the bomb where it had come to rest. Thank God, it didn't blow, for even with their hard turning, they were still within its lethal range as they passed it. When their birds slowed down to controllable taxi speed, George was able to open his canopy and jack his seat all the way to the top. After over 1 hour and 30 minutes, he was finally able to see out again. As they taxied in slowly, they glanced over and saw an amazing sight on the old grass runway. It was a huge stone cylinder being pulled by about 200 laborers. The big roller was about 16 ft in diameter by 25 ft in length. The center had been hollowed out somehow, and a large telephone-pole-sized tree trunk had been inserted in the hole to serve as an axle. A battleship-sized rope was tied around each end of the pole, and the laborers were pulling it up and down the runway. No one had any idea what this stone might have weighed. It was the Chinese answer to road graders, bulldozers, and the whole range of road building devices.

As they drifted by this amazing sight, suddenly sheer chaos erupted. Just as a stone pitched into a small brook disrupts the stream's flow, upstream as well as downstream, their unusual landing had caused a disruption to normalcy. The tower started screaming, "P-40 and P-51 on final approach to the north, pull up and go around." Their strident calls continued but the P-40 proceeded to land on the grass behind George and Bill. The P-51 pilot behind him also continued to land, but for some unexplainable reason, he landed directly on top of his leader. The P-40 was squashed into the ground, whereupon it started to shed propeller blades, landing gear, and other miscellaneous parts as it ground to a halt. The P-51 bounced high into the air after losing its landing gear and soon settled violently to the ground. It too began shedding parts and suddenly veered to the right to slide down the closest line of Chinese rope pullers. When the plane came to a stop, it left a trail 1000 by 35 ft of human hamburger. Over 100 laborers had been literally pulverized, and another 50 to 60 were crushed, maimed, or broken.

When Bill and George shut off their engines, they merely sat there for several minutes until they gathered enough strength to climb out of their planes. Then they went running to join the parade of people going out to the scene of the tragedy. There they were amazed to be met, not by grief and horror, but by 200 grinning Chinese laborers. As they walked through the carnage, many of the laborers flashed a smiling thumbs up signal and said, "Ding hao," which means "very good."

Bill and George caught a ride up to the mess hall to refuel their overtaxed bodies. Bill had once read that the mental and physical demands of 1 hour of aerobatics result in an average-sized pilot losing 7 or 8 lb of body fluids. Considering the strain they had been under, even though both of them were on the skinny side, they must have lost 15 lb each. They couldn't get enough fluids. But the strain was not yet over.

As they strolled back to the flight line, Fred and the other three P-47s were taxiing to the ramp. While they all stood there, reviewing the events of the day, they gradually became aware of the impending problem. A P-51 was on the east-west runway near them making his pretakeoff engine checks. Every time the pilot changed to another magneto position, the engine backfired and refused to run. It was obvious to all of them that this airplane would never fly. But to their amazement, the pilot gave the plane full power and tried to take off. He got it off the ground successfully but didn't have the power to climb out. On top of that, he did not use enough rudder to keep the plane going straight, and the P-51 lurched over to the left, heading straight toward them. It passed right over them and their planes but started settling to the ground into a long line of parked P-40s and P-51s. First, the straying P-51 mushed into a P-40 and removed its propeller and various engine parts. It bounced into the air and missed the next plane but squashed down into the next one in line. From there, it continued settling and literally drove through three more fighters. Fortunately, none of the planes caught fire, and the Chinese pilot scrambled out unhurt.

With that, Bill and Fred looked at each other, nodded their heads, and said, "Let's get out of here while we are still alive. The natives may decide that we are responsible for this carnage." Before they abandoned George, they charged him with finding out what happened to both of the Chinese P-51 drivers.

Several days later, George came straggling in by a wandering C-47 gooney bird. That night, they debriefed him on everything that had happened at La Hokow. He told them that the first pilot, who had wiped out the line of laborers, would only say, "I got confused." When asked why he landed on his lead P-40, he would only shrug his shoulders and say, "I don't know." The pilot who should have aborted his takeoff was slightly more articulate and replied, "I took off rather than taxiing back to the ramp because if I had brought the plane back I would have lost face." To questions such as, "But didn't you know you would never make it?" he would only say again, "But I would have lost face." Our heroes thanked their lucky stars that they were flying in an all-American fighter group and concluded that they would never understand the inscrutable Chinese.

The Jet Stream
and Bum Steers

Told by Enoch B. Stevenson, Jr.
Major, U.S.A.A.F.

Our mission was a routine escort mission to Munich. We picked up the bomber stream shortly after crossing into France. Their formation was pretty good. They were B-17s and had less trouble staying together than did the B-24s. The cloud buildup kept forcing us higher and higher until we were about 35,000 ft.

About an hour and a half after takeoff, the big friends dropped their bombs and turned for home. My thought was that Munich was a lot closer than our briefing called for, but who was I to argue with those guys? Little did I know we had ridden the jet stream to Munich and we would have to battle it all the way back to England.

We stayed with the bombers for quite some time on the return trip. One of our favorite ploys was to get a steer from ground control, so when the bombers, navigator, and all asked us for the course home, we could tell them and have one more bit of one-upmanship. So I called control but got no answer. After about 2 hours, I finally was able to raise them and complete the mission successfully.

Incidentally, on that same night when we got over the corner of France, almost to the North Sea, I began to pick up some steering calls on the radio that didn't sound right. I began to wonder about them. We were being told to turn 90° off what our course home was supposed to have been. Of course, I later realized it was the Germans trying to get us to turn the wrong way.

The entire mission of 5 hours was about one-third going and two-thirds returning. It was quite a revelation. It wasn't until years later that the term *jet stream* came into popular usage.

Break Right
and Up

Told by Joe Thompson, Jr.
Major, U.S.A.A.F.

There is one particular mission that in a way was the moment when my life was saved, but it was not of my own doing. It was after Doc Trimble had said, "That's enough, Thompson," and I had been taken off flying status. I had occasionally flown some missions after that time. But the photographic officer came on February 10 and said we needed some Merton Obliques below Mecinich, that they were crucial to the advance of the troops there, and Major Thompson was the only one who knew how to do this the way he wanted it done.

The photographic officer's name escapes me. He was an overweight guy with plenty of brain and a marvelous capacity to read photographs with his three-dimensional technique. The Merton Oblique was a photograph named for an engineer who had discovered a way to tie in the position of artillery pieces with a grid that was placed over the photographs after they were made. This permitted the artillery to fire immediately to a point on the grid and hit it the first time instead of the old over and short 200 right or 200 left, the old stuff that was standard artillery adjustment. We took the pictures at a proper altitude aimed exactly where he wanted them, and the grid was affixed in the lab where we printed pictures. Then these photographs would be given to the little guys in the Piper Cubs. They could call a six-figure coordinate on the Merton Oblique, and the first round would hit where they called.

This came as a complete surprise to the advancing Germans, and they thought it was some kind of a secret weapon we had that guided the rounds to the target by radar, but instead it was this Merton Oblique. Anyway, it was my mission, and intelligence said, "Okay, let him do it." A couple of us went out to fly this flight, and it required flying straight and level for about 3 minutes right down a road strip parallel to the German front lines and, of course, at 3000 ft right down in light flak range. Now, the danger of flying straight and level like anything else is, if you give a gunner time enough to pull his weapon through your straight flight, he can shoot you down. If there are enough of them, the chance of survival is pretty small, but they really didn't believe Mustangs would move as fast as they could move.

So what we did was come in at about 12,000 ft, making sort of a zigzag flight headed as though we were going back another way and then peel off. I peeled off from about 12,000 ft and at the last moment headed the plane exactly on the heading needed for the Merton Obliques. It was moving then, going 350 mi/h, I expect. Coming out of the dive and straight and level, I turned on the camera, kept the plane on the right heading, and noted that behind me, although I couldn't see it, was a bunch of flak bursts that were following my track and were slowly increasing in their range and getting close to my tail. The No. 2 man was to call out flak and let me know how I was doing. Those few minutes, maybe 2 or 3 minutes to do that, seemed like an eternity. But then, toward the end of the time when I was going to pull off that position, be able to jig around the sky, and keep from getting hit, I experienced an intuitive moment. It was not an eerie feeling and it was not a loud trumpet call from Lord God Almighty Himself, but it was something different.

You must understand that all aircraft turns in the European theater went to the left; you circled an airport to the left to land. If you were in a four-fighter position and there was a break, it was break left break, and everybody broke at the same time, so left-hand turns were standard. Nobody turned to the right automatically; it was always to the left, but at this particular moment, some little voice, some little indication, nothing I heard in the earphones, but something said, "Thompson, break right and up." And so when I got to the end of that Merton light grid photographic measurement, instead of turning left and pulling high, I broke right and up. At the same time, the No. 2 man called out, "Flak, flak," and I looked over to the left and exactly where I would have pulled up—they knew I'd do it—they were all aiming at the natural point that would have brought me down.

Well, we jinked our way out of that and got back in a hurry, and the pictures were great and the mission was successful. I flew only two or three artillery adjustment flights after that. The last one was on March 2. In fact, there were two that same day; one was a test hop, and then later, I went to a 240 howitzer mission and that was the last real combat mission. Anyway, I wanted to put that in sort of a windup because I had no great religious conviction at that stage of my life. I went to church and I prayed to God occasionally, but I didn't have any high-level knowledge at all. But there was that one moment when God must have had some other plans for me, and I was about to turn the wrong way. I'm convinced now, maybe it's because of memory, but that's the way it happened.

Attack on Tachikawa

Told by E. H. Bayers
Commander, U.S.N.

In February 1945, we were flying from the U.S.S. *Yorktown*. My squadron, VF-11, was flying F6F Hellcats. We were assigned to provide escort for a maximum effort strike by two other squadrons made up of 14 SBD Dauntless dive-bombers and 14 TBF Avenger torpedo bombers. The target was the Tachikawa engine plant near Tokyo.

Despite the task force doctrine of assigning a 24-plane fighter escort to a maximum strike over Tokyo, poor availability reduced the Air Group Three escort to only eight planes. To add to the difficulties, one section ran into icing conditions near the coast which limited their propeller governors to a maximum of 2100 revolutions per minute. These planes, forming the second section of a low cover division, were unable to counter any enemy fighter attacks and, consequently, joined the VB and VT squadrons, dropping their bombs on the target. These two planes scored one direct hit and one near miss from 5500 ft. The two remaining fighters in the other low section stayed with the 28 bombers and maintained a weave to furnish whatever protection they could when the anticipated fighter attack came. This attack never materialized due to the efforts of the lone high cover division.

This four-plane division, under my command, fought off an attack by 15 to 18 Japanese fighters, shot down two, probably a third, and damaged at least four others. The Japanese attack was so extensively disrupted that they were unable to get at the VB and VT squadrons. I ordered the entire division to jettison their bombs when I saw how things were shaping up, and it's just as well in view of what happened.

The story of this *Thirty Minutes over Tokyo* might well be described as "the roughest ride I ever hope to take." It was a strenuous fight for all concerned, but thanks to strict air discipline and excellent tactics, we accomplished our mission without the loss of a man. All planes landed safely aboard the carrier except Ensign Onion, who was forced to make a water landing near one of our destroyers.

Because of the weather, we had passed over the coast due east of Kasumigaura Lake and then headed around Tokyo Bay, passing approximately 30 mi north of the city. We saw numerous enemy fighters of various types in the air. Several of these made attempts to lure the escort away from the VB and VT, coming close aboard the formation singly and then turning away, offering a very juicy target. After these attempts failed, they concentrated on my division of four planes flying high cover at 17,500 ft.

They positioned three Tojos about 2000 ft above us on the port side, four Zekes in the same position on the starboard side, and four Zekes directly above us at 25,000 ft. There were six to eight scattered planes directly astern of us at the same level. When the planes astern closed for a no-deflection shot, we began our weave. Upon the commencement of it, the planes positioned on the port and starboard sides would make highside runs. Another tactic was to attack in a two-plane column, with approximately 2000 ft between planes. After the attack by the leading plane was warded off and an opposite scissors commenced, the second plane was beautifully set up for a shot. There was nothing wrong with the Japanese tactics or with their gunnery in view of the fact that three of our planes were damaged. The great weakness in the attack was that the Japanese were not aggressive enough. Numerous times, they initiated an attack only to break it off whenever a defending section came within 90° of bearing on them. The doctrine of Fighting Squadron Three, of which I had recently taken command, was a delayed (out of phase) weave. In this action, I am convinced that we would have lost at least two planes if this weave had not been employed. We used the weave exactly as laid down originally by Commander John "Jimmy" Thatch, and it worked.

We were under almost continuous attack for well over 30 minutes. The Japanese did everything possible to keep us from working our way toward the coast after the attack. At times, they succeeded in changing our course 180°. They were apparently trying to force us over the city of Tokyo, and once there, we met intense and accurate antiaircraft fire at 17,000 ft. Only one Japanese aircraft attacked us while we were in the antiaircraft sector, but they resumed the attack as soon as we were out. Although we were unable to follow any of the planes we shot at because of the necessity of maintaining the weave, the TBM turret gunners followed the battle and saw two enemy planes crash and a third dive away in flames. Several others were damaged, perhaps severely, as the number of planes attacking decreased significantly as we went on.

I was worried about Ensign Onion, but he stayed with us and maintained the weave despite the damage to his plane and his own painful wounds. It was with real relief that we left the coast and proceeded unopposed to the task force. The final strain lifted when I saw Ensign Onion make a successful water landing and climb aboard the destroyer.

Combat Missions over Northern Italy

Told by George M. Blackburn
Major, U.S.A.A.F.

It was the first week of February 1945. I was flying a P-47 Thunderbolt with the 66th Squadron of the 57th Fighter Group. Our theater of operations was northern Italy. I was leading four aircraft on a mission to knock out a bridge that was the key to the supply route from the Brenner Pass to the German Army. We destroyed it a week before, but working at night, they had rebuilt it. We made a good dive-bomb run and destroyed the bridge along with three gun emplacements. We also left about 200 yd of twisted track. Mission accomplished, we formed up and headed south. No one was hit. It had proven to be a milk run. I was glad my 76th mission was nearly over. The sun was bright and warm that day, and the sound of my big Pratt and Whitney engine was like the sound of music. It was relaxing time. The tension and anxiety of being shot at were fading fast.

As we approached the Po River, I caught a bright reflection in the water. I called to the flight and told them we were going over to check it out. As we got nearer, I was astonished. Unbelievable! Six huge barges in the middle of the day, crammed with German soldiers. "Jackpot! We are going to hit them. Let's climb for altitude." We climbed to about 7000 ft, went in stern, and circled the target. I really felt sorry for all those soldiers packed on those barges, and I thought, "What we really want to do is sink those barges because they have some 88s on deck. Knock out those big guns!" So as I banked in to start my run, I set off a few quick bursts to let them know that we meant business. We were coming in, and I hoped to God they would jump off into the water. It was my warning. It was exactly what they needed to do to save themselves. As I barreled on in, I could see that they were absolutely hypnotized. No one seemed to move. They were terrified. I had no choice. I fired my guns. As I raked across the barges, it was the closest I have ever come to a German. I was appalled at the carnage those .50 calibers could inflict on a human being. As I pulled up, I looked back, and the other three aircraft had picked out their own barge for destruction.

Finally, many of the men were seen bobbing in the water. We continued to pound away at the barges. It was incredible what eight .50 calibers hitting on a point could do to heavy metal. Before we were through, we had all six barges sinking, which of course meant the guns went down with them. I don't know what kind of casualties there were, but they paid a heavy price for trying to retreat in the daylight.

We formed up and headed back south to our field. At the debriefing, intelligence was elated; in fact, they were ecstatic over the importance of the target and the fact that it proved the Germans were in a retreating posture and we had knocked them out.

Captain Apostilou, our briefing officer, even suggested that the targets in the Po would get fewer and fewer in the daytime because of their exposure. He commented that the only Germans that would be seen in the daytime would be those with their hands over their heads. He proved to be prophetic.

As I turned to leave, Apostilou put his hand on my shoulder and said, "Blackie, I'm going to write you up for another cluster to your Air Medal and send it up to tac headquarters. I believe it will fly. Those guns would have been firing at you fellas tomorrow. It was a good show."

I looked at him and sighed. "You know, Al, I don't know how those soldiers on the ground go through hand-to-hand combat, day after day. We hit and fly away. They live with it in the worst conditions, like cold rain, mud, lousy food, and the horror of seeing and touching death. Thank God I'm a pilot. I wonder if as a dogface I could have taken it."

Al moved a step closer. "Look here, Blackie, I'm a ground pounder too, and I've got to tell you, I'm not going up there across that bomb line, not even in that B-25 parked on the flight line. No way! The ground troops want to have their feet on the ground and a medic nearby when they get shot. They don't want to fall 10,000 ft in a burning plane or float down in a chute with bullets flying by."

Carrier Crash

Told by David C. Kipp
Lieutenant Junior Grade, U.S.N.R.

I joined the Navy on July 21, 1942, in Detroit, Michigan, in the V5 program. I had just finished high school in June of that year. The Navy didn't call me for active duty until December 11, 1942, and sent me to the University of Detroit for WTS training. We had no uniforms, just the CCC uniforms, which were dark green. We were taken by bus to Pontiac, Michigan, to learn to fly Piper Cubs. Because it was January in Michigan, we soloed on skis with no brakes. We had only 8 hours of instruction.

From Detroit, the next stop was preflight school in Athens, Georgia. The most difficult requirements were learning Morse code and the obstacle course. I was assigned to Battalion 24, weighing 205 lb, and I left a trim and fit 178 lb. From Athens, we had a 3-month stint in Kokomo, Indiana, where we flew Stearmans. In Kokomo, I shared a room with Ray Killian. We became good friends and stayed together throughout the war. From Indiana, we went to Barin Field in Pensacola, Florida. In Pensacola, so many of our cadets crashed that they nicknamed it "Bloody Barin Field."

I received my commission and wings in April of 1944. However, I still had to go to Glenview Naval Air Station to qualify for aircraft carrier landings. We flew SNJs out to the converted paddle wheel, *Wolverine*. I made my eight required landings. Next I was sent to Vero Beach, Florida, for what they called "operational." I met my future plane, the huge new fighter F6F Hellcat. It was wonderful! At the end of operational, I thought I was ready for the fleet, but oh no! They sent Killian and me to Harrisburg, Pennsylvania, to fighter photo school. We flew F6Fs with huge cameras mounted in the bottom and side of our planes.

From Harrisburg, after being sent to San Diego, we were finally ready for assignment. We were to join the famous VF-6 Fighter Squadron for the third war cruise forming in Santa Rosa, California. Congressional Medal of Honor holder Butch O'Hare was its last skipper. VF-6 was sent to Hilo, Hawaii, in November 1944, to train some more. Before long, we were sent on E.T. *Collins,* a troop ship, to join the fleet at Ulithe Anchorage in the Pacific. In February of 1945, we transferred to CV-19 to the carrier V.S.S. *Hancock.*

While at Hilo, Hawaii, we had to make carrier landings with our new F6Fs. My day came and we were sent to practice one stormy day on the V.S.S. *Batan,* a Jeep carrier. We arrived at the carrier which was cruising about 1 hour off the coast of the big island, Hawaii. When it came my turn to approach the carrier deck, with my wheels and flaps down and my cockpit canopy open, I came up the approach toward the carrier ramp. The sea was very rough, making the carrier pitch heavily. When the signal officer gave the cut sign with his paddles, I pulled the throttle back. At that moment, I looked down and the carrier deck dropped at least 40 ft out from under me. We had been trained to push the stick forward if this happened, so I did. By this time, the deck had pitched back up with the sea. I hit hard on all three points, wheels, and tail, and it threw me back into the air at 70 kn.

There was only one thing to do, so I pushed the stick forward firmly and hoped for the best. My tail hook skipped and caught no wire, but I hit the barrier wire at the carrier bridge. I hit those heavy steel cables with only my wheels. It stopped me abruptly and the nose of my plane went straight down to the deck. For one long moment, I thought the plane would fall upright on its tail wheel. But it didn't. It went the other way over on its back. My two shoulder straps broke and my face went into the gun sight. After less than a minute, I could feel the tail of the plane being jacked up; then a huge white glove reached inside the cockpit and grabbed me by the chest. In one swift jerk, this white glove yanked me out of the cockpit, and the next thing I knew, I was standing upright on the deck, as the wheels of my plane were still spinning.

The hospital corpsmen were there by me and wanted me to lie down in the wire stretcher. I said, "No, thank you. I am just fine." They said, "Please lay down in the basket. The captain is watching and your face is all bloody." I said "No" again and started for the bridge door. I got just inside the door when my knees gave out. I then laid down in their basket and they took me to the ship's hospital. At the hospital, the young doctor told me he just finished training at Johns Hopkins, and if I would let him take his time stitching me up, taking very small stitches, I would not have a noticeable scar over my eye. I said "okay" and he took 11 stitches over my right eye. To this day, the scar is hard to see. This, too, is the same reason I didn't get a Purple Heart.

Armed Reconnaissance: Southwest Belgium

Told by Johnnie B. Corbitt
Captain, U.S.A.A.F.

I had that "uneasy" feeling. Not only had I somehow lost my lucky pilot class 43-E ring, but a friend had jokingly told me not to bring that old plane back because it had not been acting right lately.

After making several attacks on German transport units, I realized that I had lost radio contact with my flight leader. I thought the radio had been damaged by flak, as I had flown through one large burst that had exploded right in front of my nose. I followed along with the mission and we soon spotted a long line of horse-drawn vehicles under the ever-present double row of Lombardy poplars. We attacked and made pass after pass in a virtual traffic pattern. On one pass, after firing a long burst, moving from target to target by kicking the rudder, as I tried to pull up, the plane responded very slowly and crashed into the crown of one of the trees. The plane came out of the tree almost upside down just about treetop level. I reacted instinctively with full throttle, full left rudder, and stick left and back. To my amazement, the plane righted itself and was straight and level, barely above the ground. While this was going on, I thought, "My parents will be so sad," and I felt very badly about the ugly attitude I had used to badger them into giving permission for my entry into the air corps. I also thought that the Germans would be watching and laughing at my predicament after what I had just done to them. What's more, I thought I was floating in the air above all this watching the plane crash and roll into a ball of aluminum with flames coming out of the rolling ball. It was a very weird experience, just as vivid today as it was then.

After regaining control, I took stock of what just happened. The windshield was covered with green steaming "juice" from the tree. A large limb was lodged in the left wing against the main spar inboard from the guns. The wings were battered from tip to tip, and the horizontal stabilizer and elevator on the right side were sheared off near the fuselage making the plane very nose heavy as the trim was destroyed. I quickly had to wrap my legs around the stick to hold the airplane in the air. Although I had no radio contact, I babbled something to the effect that I was going in, meaning back to base, into the useless microphone. The flight leader heard me and thought I was crashing. I was a little excited and probably not too understandable.

I turned and climbed slowly to about 7500 ft and flew in cloud cover to avoid detection. As I neared the Seine River, I ran out of clouds and was flying in the clear. Suddenly, an FW-190 flew up and across my path. I started getting ready to bail out since I was helpless. Out of nowhere, a P-47 appeared on the tail of the FW and they kept going. So did I.

As I neared the beachhead, I tested the aircraft to see if I could safely land it. I was flying at 195 mi/h at full throttle. As I slowed to land, it stalled and rolled at 190 mi/h. I recalled that our CO had said, "Boys, don't kill yourselves trying to save a beat up old airplane. It takes 20 years to make a pilot, and they make these things every day." I took him at his word.

I disconnected all the cords and harnesses, opened the canopy, rolled the plane over, and was catapulted out due to the nose-heavy condition. At first, I seemed to be sitting in the air, but then I started tumbling. I reached for the D-ring and couldn't find it. I remember that after I opened my eyes I saw some object sail off and away from me. I thought the jagged metal from the stabilizer had cut my chute away. It turns out the object was my goggles that were torn off when I hit the airstream. We had just been issued new English-type parachutes, and the D-ring had slipped around under my left arm. I found it and jerked mightily. The big, beautiful parachute blossomed above my head. I reached for the shroud lines and my head jerked to one side. I had the earphone jack clamped between my thumb and finger, and the D-ring was clamped in my right hand. I had to literally pry them loose and put the D-ring in my knee pocket.

I took stock and heard the most awe inspiring silence. *No sound!* I heard a plane buzz me two or three times, but I never saw it until I was on the ground. I landed in the only open field around. It was a square field surrounded by hedgerows and apple orchards. Suddenly, French people came from every direction. We couldn't communicate, but one of them handed me a small, dainty glass stem while another filled it from a long-necked green bottle. I held it for a moment as my hand started shaking. I downed the clear liquid fire in one gulp and reached for a refill. I emptied my pockets of Lucky Strikes, which thrilled them. They wanted my parachute, but I wasn't about to give that up. It turned out that it had a rip in the canopy and could not be reused.

About that time, a Jeep roared up from the nearby air base. They had seen me bail out and came looking for me. They took me to their squadron operations tent. The operations officer asked, "What happened? We heard the engine cut out and saw you jump." I told him, and he said, "Makes you feel like a damn fool, doesn't it?" "Yes, sir," I answered. "I know," he continued. "The same thing happened to me last week." I was sure glad to get off the MIA list when I got back to my unit, the 493d Fighter Squadron.

The next day, our group moved to Villacoublay Airdrome in Paris. As my flight swooped down low to buzz our former home farewell, I stayed up a couple of hundred feet higher.

How It Ended

Told by Roy D. Simmons, Jr.
Captain, U.S.A.A.F.

March 1, 1945, flying from airstrip A-95 located south of Nancy, France, proved to be my final combat mission. I was briefed to conduct a reconnaissance mission of the retreating German forces from in and around the Battle of the Bulge area. The Germans were certainly retreating, yet they possessed an abundance of groundfire power, ready to be thrown at anything, whether in the air or on the ground.

Before I relate the events of this mission, perhaps I should first discuss developments of the previous 3 months. On December 2, 1944, the commander appointed me as the squadron's engineering officer. To quote him, "Roy, you now have 106 combat missions. You can go home anytime you wish; however, I need an engineering officer and I would like for you to stay a little while longer." Sounded great to me, yet he was not through. "I am also grounding you from combat. However, this grounding in no way affects required test flights." Needless to say, I was pleased with the new duties, yet disappointed with the grounding decision.

The winter of 1944–1945 was tough in France, and maintaining the in-commission rate of aircraft became a real problem. Flight testing of aircraft became increasingly necessary, and being as close to the front lines as we were also required the testing of the aircraft guns. I'll leave this comment to your own imagination, as a considerable quantity of ammunition was expended, even by an engineering officer.

Following weeks of badgering, the commander reluctantly agreed to let me fly combat, but only one mission each week. I was elated with this change of heart, so on February 8, 1945, I was back to flying combat. Now, my story.

I climbed into *Flying Jenny II*, happy to be back to combat flying and feeling confident and comfortable in this beautiful plane. *Flying Jenny I* was shot down on Christmas Day 1944 during the Battle of the Bulge. The pilot was killed. On our way toward the target area, climbing through the cloud cover, observing my wingman neatly tucked in on my right wing, I hoped the target area would be clear. But it was not to be. The area was covered with patches of low-level clouds and fog. I had been instructed to get in close for a visual and also take as many photos as possible since Army headquarters was interested in the quantity of rolling stock the Germans still had operational.

Dropping down through a break in the clouds, it immediately became apparent this mission would be "heads-up flying" in every respect. As if waiting for us, the groundfire became intense. It seemed even worse with the flashes of guns firing in this dark weather environment. I instructed my wingman to be alert and to call out "hot spots," while I concentrated on the assigned mission.

To visually spot vehicles, troops, fuel dumps, bivouac areas, and so forth became difficult, yet the photos taken turned out quite well. The weather was a major factor. Low clouds and fog eventually forced us to leave the area and head for the bomb line; not a minor consideration was the intense and accurate groundfire. We began a climb through the overcast, breaking out into the clear at about 9000 ft. Fortunately, neither of us had experienced any aircraft damage. The intense groundfire, lousy weather, and the difficulty in observing the area had resulted in this being a taxing mission, and we were not home yet.

I called for a directional finding (DF) steer to our airfield. A clear, crisp voice requested I count slowly from 1 to 10, which I did, and immediately received a heading of 20° to my right, which seemed a little strange. Yet, being over a solid overcast, I didn't question the controller's instructions. Shortly, I again requested a steer, and as before, a heading of 20° change to the right was given. This was also strange. Then, all of a sudden, it dawned on me. This was a German tracking station that was slowly bringing me back over enemy territory. So without further ado, I disregarded these instructions and returned to my original heading. A station near our airfield responded to my call for a DF steer, gave me a heading, and cleared us for a descent through the overcast to an open area. The ceiling was now 2000 ft. Airstrip A-95 never looked so good as we came in and landed on good old terra firma.

I was rather pleased with the results of the mission in spite of the adverse conditions under which it had been flown. Yet, I had an uncomfortable feeling, a feeling that somehow I had not been in complete control. An uneasiness had come over me. I contacted the squadron commander and requested the "one mission a week" restriction be lifted. I explained to him my feelings regarding the recent mission and that I needed more combat flying. Infrequent combat flying would get me killed. His response was, "One mission a week, no more." I then requested a 30-day leave of absence, to go home, and then return to the squadron. Again, his response was, "No." Getting nowhere, I then said, "Send me home PCS [permanent change of station]," to which he agreed, saying, "Roy, you have done enough. It's time for you to go home."

So ended my World War II combat flying, having flown 110 missions. In addition, I had held the positions of squadron operations officer and engineering officer. One of the most difficult things I had to do was bid farewell to my crew chief, Corporal Kujawa, and to *Flying Jenny II,* as we had become quite a team.

First Mission

Told by Clifford J. Harrison, Jr.
First Lieutenant, U.S.A.A.F.

It was March 1945, and after crossing the Atlantic in a slow convoy, I had at last joined the 526th Fighter Squadron as a replacement pilot. It seemed like a long time since that day in April 1943, when my senior class (happy to be getting half a day off from school) saw me off at the train station. I had enlisted in the Army Air Corps Reserve shortly before my 18th birthday, which was in February. This was necessary to get service of choice. I was told that I would not be called up before I graduated in June.

The 86th Fighter Group was in Italy when I started my journey across the ocean. By the time I arrived in Naples, the group had moved to Tantonville, France, a small town south of Nancy.

The day after joining my squadron, I flew a short transition flight to be sure I could still fly a P-47. To me, the most unusual thing about my first flight overseas was the wire mesh airstrip on the side of the slope. We took off downhill and landed uphill regardless of the direction of the wind.

Very early in the morning, I headed out on my first mission flying the wing of Captain Eli Stencich, my flight leader. Prior to the flight, we had a short briefing by our intelligence officer which meant little to me. I do remember a red line across the map that was referred to as the "bomb line."

Once airborne, it did not take me long to figure out that we had crossed the bomb line when I saw a number of black bursts in the sky all around us. Our flight leader's voice came in on the radio, "Package Gold Flight-Jink." At this, the other three airplanes in the flight started making all kinds of erratic maneuvers, and of course, I joined them. Before too long, we were close to Mannheim, which was our target area. We were about 10,000 or 12,000 ft and below us was a marshaling yard with a number of locomotives. We were carrying a 250-lb general purpose bomb under each wing, and it was obvious we were going to dive-bomb the trains.

My flight leader rolled over and began his dive-bombing run almost straight down. I will say that his rollover and the steepness of his dive took me by surprise. It was a much more decisive maneuver than I was taught in my P-47 training back in the States. In any event, I did the same thing. Then I realized that my wing blocked my view of the leader. I thought I would see him when he pulled out of his dive and that I should wait until he pulled out before I released my bombs. It doesn't take long to get to the ground when you are diving straight down in a P-47. The marshaling yard and the trains were getting very close. Quite a bit of flak was being thrown up at us, but the thing on my mind was when to pull out of the dive. It seemed that I was flying right into the ground when I decided it was time to release my bombs and pull out. I found that one hand pulling on the stick was not enough. I took both hands and pulled back with all the strength I had. Of course, I blacked myself out and when I recovered consciousness I was going straight up. I saw the other P-47s in my flight and happily rejoined them.

After I landed, my crew chief told me my plane had been hit and that he found three holes in the wing. He told me that a pilot got 2 oz of whiskey for each hole, and he would see that they were duly reported. I guess the fact that I was under the legal drinking age did not matter.

I was told I was to fly another mission that afternoon. At least I now knew what to expect on a dive-bombing run. I would not be surprised again.

Rescue of a Downed SB2C Pilot: The Battle of Kagoshima Bay

Told by Roland H. Baker, Jr.
Ensign, U.S.N.R.

On March 29, 1945, the U.S.S. *Hancock* CV-19 launched 10 SB2C Helldiver dive-bombers. Each aircraft was loaded with two 1000-lb SAP bombs. They were accompanied by a division of F4U Corsairs led by Air Group Six Commander Henry "Hank" Miller. Failing to find their initial target, the enemy fleet, the flight headed for their secondary target on Kagoshima Bay. The bay was a fat finger of water pointing toward the heart of Kyushu. The mouth of the bay spilled into the East China Sea. The entire area was defended by strategically located airfields and by a heavy concentration of antiaircraft guns.

Over the naval air station at the head of the bay, Lieutenant "Slim" Sommerville pulled out of his dive while taking evasive action to escape antiaircraft fire. Pulling up into the overcast at 2000 ft, he felt a terrific jolt. The tail of his plane was cut off in a midair collision with another dive-bomber. Sommerville bailed out at 800 ft. His rear seat gunner was dead. Overhead, Commander Miller saw the collision and watched the single "chute" leave the stricken plane. He waited until he saw the inflated life raft and then radioed for a rescue attempt. He ordered the fighter division to provide cover to the downed pilot as long as fuel allowed. His radio message was picked up by Lieutenant Robert Klinger who was flying combat air patrol over the task force 150 mi away at sea. The division of F6F Hellcats consisted of Klinger, the division leader, his wingman Ensign "Willie" Moeller, Lieutenant Junior Grade Louis Davis, and me. We were ordered to proceed immediately to Kagoshima Bay. The plan was to protect Slim until a rescue could be mounted. The four Hellcats sped to the bay and relieved the fuel-starved Corsairs on station.

Davis and I were sent above the broken cloud layer to watch for enemy aircraft while Klinger and Moeller covered the downed pilot. The Japanese knew a pilot was down and had reason to know only four of us were protecting him. Eight Navy Zeros (code name Zeke), in tight formation, hit us from above. We spotted them quickly and called for Klinger to come up. One of the Japanese peeled off and started a strafing run at the raft. He never pulled out of his dive as Davis got on his tail and finished him with an accurate burst. Another Zeke flashed in front of me presenting a perfect shot. I then looked over my shoulder and saw a Zeke diving down on me. Wrapping my turn, I climbed to meet him head on, both of us firing at the rapidly closing adversary. As the Zeke passed by, I turned hard over and observed it go straight into the bay. Then I let my guard down momentarily as I watched Davis's Hellcat hit the water. He did not get out. I now found a Japanese on my tail and sustained a few hits. He was gone as quickly as he appeared, as Klinger shot him off my tail just in time. Later, Moeller returned the favor by shooting one off Klinger's tail. Seven Japanese aircraft were in the bay before it was over.

The rescue group from the *Hancock* arrived with two OS2U Kingfisher seaplanes from the heavy cruisers. Two more Japanese were sighted and destroyed. One of the escorting

Hellcats was hit by antiaircraft fire and made a water landing next to Sommerville. The seaplanes taxied over and picked up the downed pilots. A thorough search failed to locate Davis, whose courageous attack in the face of great odds cost him his life. Then we got the hell out of there.

Upon arriving at the *Hancock*, I was unable to lower my landing gear due to the damage caused by the Zero that had been on my tail. I ditched alongside the destroyer *Stemble*, returning to the *Hancock* a few days later.

Kamikazes

Told by Hensley Williams
Major, U.S. Marine Corps

By the time of the battle for Okinawa, the Japanese air forces were quite weak, as a rule, and there were few days during which they seriously challenged us in aerial combat for surpremacy or control of the air. But there were many times, mostly during night or bad weather, when they sent their kamikaze (suicide) planes and pilots against us. At times, they did great damage to some of our ships and some damage to our land forces as well. The majority of our missions were in close air support of our ground and naval forces by dive-bombing, rocketing, and machine-gun strafing. I spent many days flying circles around our naval picket ships which had been badly hurt by the kamikazes. Finally, it paid off for me and my wingman (a two-plane tactical section) when we spotted two kamikazes in bad weather and low visibility. They were flying close to the ocean surface and heading for our picket ship. It was almost like shooting fish in a rain barrel. However, at least I came out of the war with one aerial victory to my credit. That is satisfaction of a sort to a fighter pilot who really never engaged in an aerial dogfight with the enemy. But with 52 combat missions to my credit, I could at least feel I had done my part in the war.

Kamikaze

Told by David C. Kipp
Lieutenant Junior Grade, U.S.N.R.

I would like to tell you of one incident on my tour that stands out in my mind. I was with Air Group Six, aboard the carrier *Hancock* (CV-19). It happened on April 7, 1945.

We were about 100 mi off the coast of Okinawa. My division of four F6F Hellcats, led by Lieutenant Wickham, had just completed a bombing and strafing mission. No Japanese fighters had been encountered. We returned to our carrier at about 1000. The carrier turned into the wind and announced they were ready to land us aboard. We landed in order. There were no wave-offs, but it was here that things changed dramatically. As I landed my fighter and looked down the deck, I noticed something very different. There was a large bulge in what should have been a flat, level deck. I also noticed that the men directing my taxi out of the last hook wire were making me stay way over to the right-hand side of the ship. As I taxied toward the bridge, I very soon noticed a huge hole in the deck on my left side still smoking and charred from a fire.

I parked my plane, got out, and asked the men what had happened. They told me that 2 hours ago a Japanese kamikaze had dived out of the clouds and straight through our flight deck. I walked back to the smoking hole, about 50 by 75 ft, and looked down. The six to eight men stationed to the left of the 20-mm gun station lay there black and lifeless. No one, as yet, had had time to remove the bodies.

The next thing I wanted to do was to find Mac, my plane captain, who had stuffed me in my plane only a few hours ago. I finally found Mac's best buddy who was also a plane captain. He told me that when the kamikaze came down, there were about 25 men on the deck by the bridge. He said when they saw they were going to be hit, they had to make a fast decision to run forward or aft. All the men that ran forward were alive. Those that ran aft made the wrong choice and were killed. All that was ever found of Mac was one shoe. Because I had known Mac so well, my commanding officer asked me to write a letter to his mother and father explaining what had happened. That was one of the hardest letters I ever had to write.

We lost 65 men from the bombs and the fire. We had no water or electricity on the ship that night. The next morning, we left the fleet with two destroyers. We went to Ulithi Anchorage near the Philippines. We picked up some airplanes there that needed repairs and took off for Pearl Harbor for ship repairs. I was 22 years old and Mac was 19. He would have been 72 today if he had run forward rather than aft. I think of Mac often.

A Good Day

Told by O. T. Ridley
First Lieutenant, U.S.A.A.F.

The day was relatively clear with high clouds back to the east and some lower middle clouds to the southeast. I had just completed my third circuit of the Prague Airdrome in pursuit of a Messerschmitt 262. The time was about noon on April 18, 1945.

Captain Charles Weaver and I had flown into the area of Prague, Czechoslovakia, as spares for the 362d Squadron. The mission was to intercept the German jets after they hit the bomber stream and were returning to home base, low on fuel and ammunition. There were no aborts. Captain Weaver and I were free to hunt the area. In the process, we engaged two ME-262s.

The antiaircraft fire around the airdrome was heavy, but their gunners were consistently shooting behind me. I had been getting strikes on the ME-262 and felt that another circuit or so might prove fruitful. The speed at which the ME-262 was flying allowed me to get only three shots at it during a complete circuit of the airdrome. I was flying the P-51 balls out.

During the northerly portion of my turn to the left, I was hit by antiaircraft fire on the left side of the fuselage in the vicinity of the No. 5 exhaust stack. The P-51 lost power, so I retarded the throttle in an attempt to coax more power from the engine. The engine made a slushing sound similar to that heard when clearing a stopped drain with a rubber plunger. Fire immediately started in the area of the No. 5 exhaust stack. I turned the airplane west and started to climb.

I called Captain Weaver, told him I had been hit, and was bailing out. I asked that he write my parents with details. I jettisoned the canopy, disconnected oxygen and radio connection, as I recalled the dash one procedures recommended for bail out. One was to slow the airplane and dive out toward the trailing edge of the left wing. I stood up, but there was considerable fire and molten aluminum coming back from the engine. I sat down, reconsidered my bail out procedure, rolled the airplane over, and dropped out. I did not see the P-51 again. It was like losing an old friend. I had inherited the P-51 (serial number 44-14789) from Major John England. The airplane had its own personality and peculiarities, together with a colorful history, including over 25 kills. This was probably the demise the airplane preferred.

I got out of the airplane at about 3000 ft. When I stabilized, I was in a faceup position. I could see the horizon between my boots, with clear blue sky above. My departure from the airplane was in such disarray that I had difficulty locating the parachute rip cord. After what seemed like an unusually long time, I found and pulled the rip cord. The parachute opening came with quite a shock. I lost helmet and oxygen mask and later found that my knife and personal survival gear had gone through the bottom of my lower right flying suit pocket. After the parachute opening, I could hear voices and a sawmill in operation. The voices seemed to be those of young people or women. We had been told by previous evadees at squadron briefings that one should avoid the Hitler Youth at all costs.

I swung in the parachute through two oscillations and was on the ground at the third. I was in a wooded area with some undergrowth all around. No bones were broken. I collected my parachute, rolled it into a neat ball, placed it under what I considered a dense bush, covered it with brush, and departed on the run in what I thought was a westerly direction. Time passed. The trees and general area began to look familiar. I had run in a circle. Previous evadees had briefed that this was likely to occur. I regrouped, discussed the matter with a more composed self, and proceeded in a deliberate walk westward.

I walked for about an hour. The woods were quiet; the trees were conifers of moderate height, still with some under-growth. The mid-April temperature was not uncomfortably cold. In what seemed the far distance, I heard someone call my name. The call was "Herr Reedleey," repeated three or so times. It occurred to me that maybe I was hearing things and then maybe not. In any case, it seemed appropriate to move on. At 1600, I had been walking for about 3 hours. I was in a heavy dense wood with much undergrowth. It seemed a good place to hide and wait for nightfall.

I retired under, in, and among a small group of conifers with undergrowth. I took off my Mae West, adjusted my back against a small tree, and proceeded to assess my situation. The bottom right pocket of my flying suit was torn open. My map of Europe, folding knife, and other personal items one might need were gone. I opened my issued escape kit. Inside was a nice colorful silk map of Europe, my three pictures, a compass arrangement that consisted of two buttons that could be used on clothing, fishhooks and line, a nutrition bar (chocolate), and some first-aid equipment. I took off all insignia, dug a small hole, and buried it in the ground. I took my dog tags, taped them together, and then quietly checked my pistol. I checked again for injuries; all was in good working order.

I figured I was about 20 mi west of Prague and with some luck I could walk to the American lines. I looked closely at the attractive silk map of Europe; the eastern edge stopped at Pilsen. I was off the map to the east. My map study was inter-rupted by what sounded like someone coming slowly through the woods. The sounds came closer. I eased as far back into the brush as possible. I could hear the person humming. I slowed my breathing but could do little to slow the beating of my heart. The heartbeats had to be audible. The soldier was very close by; the black boots went by my bush clump. I could have reached out and touched the boots. Gradually the sounds faded and I settled back to planning. I tried to memorize the towns at the edge of the map east of Pilsen in the hope that I was further from Prague than calculated. About 1700, I had moved farther west. After about 1 hour, I could hear voices in the distance; a few yards further and I could see a road. I decided to backtrack a quarter mile and wait for darkness.

Twilight came and then darkness at about 1830. I moved close to the road where I could see people walking. A good number of them carried packages. They nodded as they passed one another and said, "Dobry noc" (good night). As it got darker, I took off my brown flying suit, rolled it up to make a brown bundle that included my Mae West, flying jacket, and pistol. I was wearing a wool olive drab shirt and trousers with no insignia. I put the package under my arm. The road was clear. The briefers had said, "One must be bold." I stepped out on the road and headed west, hoping I looked like a native.

Dive-Bombing
Along the Rhine

Told by Johnnie B. Corbitt
Captain, U.S.A.A.F.

I was flying with Jake Cooper on this mission. The overcast was almost solid but cleared briefly over the target as predicted by George, our amazingly accurate weather officer.

After completing the mission in very heavy flak, we started back toward base. The weather was much worse by this time, and we were flying strictly on instruments. Jake's eyes were glued to the instruments and mine were glued to Jake's wing. After proceeding in this manner for what seemed like an eternity, I tried to contact Jake to inform him that I was getting dangerously low on fuel. I received no answer or response of any kind from Jake. I tried radio, hand signals, wing signals, everything I could think of but got no response. Finally, as my fuel gauges reached zero, I prepared to bail out as soon as the engine quit. Almost miraculously, a large hole appeared in the cloud cover and directly under this hole was an airfield covered with P-47s. The field was located near Metz in northeastern France.

I pulled up in front of and above Jake, and this time he saw me. I signaled toward the field and started descending immediately, fully expecting the engine to quit at any moment. A C-47 was taxiing up the middle of the runway for takeoff. I was not about to pull up and go around, and so was holding off landing until I cleared him. He didn't know my intentions or if I even saw him, so he started frantically flashing all of his lights and weaving the plane to get my attention. I barely cleared him and dropped the plane on the last half of the 3000-ft runway just behind him. I'm sure he thought he had a very close call. Jake landed shortly and we taxied to the parking area. I had to explain my unorthodox landing to the operations officer.

They fed us, put us up for the night, and later told me I had plenty of fuel. Flak had cut a wire from the generator, and as power was used up, I lost the radio and the fuel gauges slowly began to fail and dropped toward empty.

They repaired the damaged wire, and the next morning, we had a short 35-minute flight back to St. Trond, Belgium. We were, needless to say, happy to be home and removed from the MIA list yet again.

The Chocolate Bar

Told by William S. Miller
Captain, U.S.A.A.F

It was a warm spring morning in 1945 when Bill and his other three pilots took off on an armed reconnaissance mission to northeast China. An armed recon mission was Bill's favorite kind because it meant that he could plan all the where, when, which way, how long, what ordnance, and so forth. In short, this kind of mission meant "have guns, will travel."

They were flying the big P-47 Thunderbolt fighter. This aircraft had been built for high-altitude interception of enemy fighters, and it was superb in this role. The plane's invulnerability to groundfire and its eight .50-caliber machine guns also made it an excellent low-altitude fighter bomber, ground strafer, and general purpose fighter. It was powered by a huge radial engine that developed up to 2800 hp in short spurts. Each of the eight guns carried 425 rounds and fired 800 to 900 rounds per minute. Thus, one could fire all the ammunition in under 30 seconds. However, they didn't shoot the guns this way; a 30-second burst would have burnt out the gun barrels.

Bill was feeling especially bloodthirsty today and was looking for Japanese to kill. Last month, his best friend had been shot down over Hankow and they had just received information that the Japanese had tied a rope around his neck and dragged his body all around the city in order to prove their invulnerability.

Bill, who would be 22 in December, had now been overseas for 25 months, flying the P-39s in Panama when the Japanese were expected to hit the canal and flying the P-39 again in Italy where they did convoy patrol over the inbound ocean shipping. Now he had been in China for over a year. He had just made captain earlier this month and was the flight commander of D flight in the 91st Fighter Squadron. He had a total of 62 combat missions but he was about to get discouraged about ever finding any airplanes to shoot down. Flights he led had shot up over 40 locomotives, numerous trucks, a few tanks and troop emplacements, and had dive and skip bombed numerous rail yards, bridges, and airfields. But he wanted to be an ace so badly he could taste it. He was toying with the idea of requesting a transfer to a long-range P-51 or P-47 unit in the Pacific when he completed 75 combat missions in China. Those Pacific long-range guys were flying fighter sweeps over Japan and were shooting down Japanese planes like swatting flies. Besides, Bill was becoming increasingly aware that their flight surgeon, Dr. Max Salvater, was starting to keep a good eye on him and he was also well aware that some of the junior guys in the squadron were starting to wish that Bill and his contemporaries would go home and quit hogging the good missions and the captain slots.

In order to find some Japanese planes today, Bill had their ground crews load his flight with a 75-gal belly tank on each wing rather than the normal one on the belly. This would give them 520 gal of fuel rather than 445 and would allow them to fly for about 7.5 hours rather than just over 5. In this way, they could work the railway and two nearby airfields on the main line to Peking. He also intended to swing to the south after checking on those airfields and go another 50 mi southeast of their normal reconnaissance area behind the big Yellow River bend. There was a huge bridge spanning the Yellow River near Kaifeng that he wanted to look over. The bomber guys had consistently been unable to wipe out this structure which U.S. engineers had built in the 1930s. The bridge carried all of the rail traffic and much of the road traffic heading north or south in central China, and Bill wanted to decide if a flight of eight to twelve P-47s might be able to destroy it. At least the fighters would be able to hit the bridge with their 1000-lb bombs, which was more than one could say about the BUFs (big ugly fellows).

As Bill and his flight leveled out over Sian and headed to the east, they didn't know—and most of them would not have cared much—that they were flying over the ancient capital of China. Never mind. Most of the Chinese natives didn't know either that Sian was the capital of the Emperor Qin who successfully united China and ruled the Chin Dynasty from 221–207 B.C. They also had no idea that Emperor Qin had been buried with his famous 6000-man terra-cotta army near the Wei River right below where they were flying.

Today Bill intended to conserve ammunition and fuel until they got into the virgin territory. So, when they found their first railroad locomotive, only two of them strafed it instead of swarming all four onto it. While two were down shooting, the other two circled lazily at 8000 ft. They took turns doing this until they left their normal target area.

They were flying over an area that much resembled the hilly terrain in the southeastern United States. In fact, they were operating at about the same latitude as Bill's native Tennessee and they were almost exactly halfway around the earth from it. If one could have overlaid a U.S. map over this area, their operating base at Sian would have been near Little Rock. They would hit the Japanese at the equivalent of Memphis and would follow a rail line a couple of hundred miles up to the equivalent of Louisville, Kentucky. At Taiyuan, they would then take another rail line about 250 mi to the northeast past Wheeling, West Virginia, and into the Shenandoah Valley. Near there, they would work their way southwesterly until they struck a main north-south rail line in mid-North Carolina. Next they would follow this line to the equivalent of Greenville, South Carolina, at which time they would head westerly toward Chattanooga. After that, they would follow the Whang Ho River to about Asheville, North Carolina, to check the bridge at Kaifeng. Finally, they would go westerly and depart enemy lines "near the overland

Memphis." They would cover well over 1600 mi on their little jaunt today.

Bill had drawn the traditional course lines to and from the bend at the Yellow River where the first enemy airfield and the rail line started. It was about 85 mi to this line, which was called the "bomb line." After they crossed the enemy lines, Bill used his own system of navigating. On the scale of maps they used, when he curled his left thumb next to his forefinger, the distance to the end of his finger was 50 mi. If they were just looking for targets, it took 12.5 minutes to cover that distance, and if they allowed one strafing attack for each 50 mi, which was the average, it took 15 minutes to go that far. For directions, Bill would point his finger toward the next checkpoint and estimate the next heading. It sounds primitive but it was as accurate as the maps of China were in those days. Bill would carefully record on his map, in pencil, the time over each checkpoint or major terrain feature and would also annotate the time and results of attacks on each target. As a safety feature, in case he were shot down, he would call out their position over each significant checkpoint to other members of the flight so they would be able to find their way home.

When they passed out of their normal operating area at Sin-Chia-Chuang toward Peking, they had been airborne for 2.5 hours and had destroyed three locomotives. During the next 30 minutes, they found and shot up three more trains. The two airfields they passed over were disappointingly bare, however, so they had to content themselves with shooting up the hangars and the operations buildings.

After they checked out the last airfield at Sin-Chia-Chuang, Bill was sorely tempted to go on toward Peking. They all noticed that they were getting out of the hills and mountains, and the coastal plain was starting. He surely would like to see that city and catch a glimpse of the China coast. He measured the distance and found it was about 220 mi to Peking. If they dropped their extra fuel tanks and climbed to 20,000 ft, they would be able to cruise at about 360 mi/h. It would take them about 35 minutes downwind to the coast and about 40 more back to where they were. Fuel would be no problem and they might be able to flush out some Japanese fighters. On the other hand, fuel tanks were in short supply and they were expected to bring them back if humanly possible, and he had not included Peking on his mission plan. Bill debated the pros and cons with himself. He knew that the rest of the flight would want to go on, but concluded that they were not getting paid for joyriding. Reluctantly, he decided to stick to the original game plan.

After making that decision, they pulled up to 10,000 ft to conserve fuel and cut across the hypotenuse of the triangle heading southwesterly toward the main north-south rail line. After 45 minutes, they intercepted that line and were back in familiar territory. They cruised down this line to the south and found two more trains to work over. Score: eight locomotives. They normally didn't bother with the boxcars or flatcars on the trains unless they were troop trains or had fuel tanks or other obvious military targets. As powerful as four good bursts of .50 caliber were, they didn't do much damage to rice, clothing, or machinery.

To attack a train, they normally tried to find it out in the country, away from antiaircraft guns. Then they would get into a loose follow-the-leader formation and aim for the cab of the locomotive. In normal situations, after they had been working over a railroad for an hour or so, the alert system went up and the Japanese parked the locomotives in tall revetments that were built in the rail yards of the towns and villages. In this case, it was a little harder and the cost of living went up a bit. One could then only attack the engine from straight up or down the track. The crafty Japanese had targeted these areas with all the machine-gun, cannon, and even handheld rifles that they could muster. The groundfire didn't really worry Bill and his friends very much. One guy from this group had been killed rooting out trains from these revetments, and Bill had a plane shot up so badly it was totaled. As a matter of fact, Bill and his flight felt as if they were draft dodgers if they came back home without a few holes in their planes.

When they found the next locomotive, the other guys—
Bill's wingman George Frieze, Roy Holton, and Luther
Emreigh—started reporting that they only had four or five
guns firing and that they were close to running out of ammunition. Bill still had all eight shooting for he was a better
gunner and required shorter bursts to get the job done. Since
Bill had his armorer load a string of 10 tracer bullets when
he had 35 rounds per gun left, if he saw a steady stream of
tracers, he knew he was about out and it was time to head
for home. He did not relish having to fly over several hundred miles of enemy territory with no way to defend himself.
None of Bill's squadron mates ever adopted this safety pre-
caution, and he could never understand why. Bill could
always find something to shoot at with the remaining 25 or
30 rounds in each gun, just before he left enemy territory. As
often as not, it was the lonely Japanese sentinel, standing in
his Buckingham Palace sentry box, who was stationed every
mile or so along the railroads. Additionally, there was an airfield right at the Yellow River bend whose buildings could
always use a good strong burst if he passed that way.

Anyway, they now had been out about 5 hours and were
at least 2 hours from home, so it was time to start heading
for the ranch. They were at the Chattanooga equivalent on
the U.S. overlay of China and in 30 minutes they would
reach the bridge at Kaifeng. Soon they intersected the river
and veered 40 mi to the east of the bridge. En route to the
bridge, Bill had them spread out and relax at 10,000 ft.
They even took off the necessary but uncomfortable oxygen
masks, which also contained their radio transmitters, and let
them hang by the strap, movie hero style. This allowed their
faces to return to normal, and they were even able to smoke
a cigarette and have a drink of water from their web belt
mounted canteens.

They also wore on their web belts a survival knife, a first-
aid pouch, and their service .45-caliber semiautomatic pistol
plus two extra seven-round magazines. Since it was such
a hot day, their A-2 jackets were stowed at the right rear of
the seat. Bill had reminded them to make sure they grabbed

the jacket if they had to bail out. Bill's jacket also held a .25-caliber Mauser semiautomatic pistol, liberated from a bawdy house in Italy for a carton of Luckies. Other pockets contained silk escape maps, which one could swallow if need be, depicting the land they were flying over. Also included were two flags, one of Nationalist China which said, "This foreign devil is a friend of your country who is flying in support of your Army troops. If he is shot down and needs help, you will be handsomely rewarded if you help him." Or at least it said words to that effect. Interestingly, there was not then—and probably still is not—a single Chinese word for *foreigner.* If you weren't Chinese, you were a "foreign devil." The other flag was the U.S. flag, ringed with words that said the same thing in other oriental dialects such as Burmese, Thai, Indian, and Vietnamese. In the movies, the pilots always wore these flags on the back of their jackets. However, Bill and most of his friends felt that the red, white, and blue colors were too conspicuous. In case one tried to escape from a pursuing squad of angry Japanese, one does not want a convenient bull's-eye in the middle of one's back.

The parachutes that fighter pilots wore had the silk of the canopy in a seat pack upon which they sat. The backs of the parachutes were padded with a 2-in thick cushion of foam rubber for pilot comfort, but the foam had cutouts in it for the storage of other survival material. On top, between the pilot's shoulder blades, was either a 13-in machete or an 11-in Bowie knife, pilot's option. Bill opted for the Bowie since they were too far north for bamboo forests and jungles. Next, down the middle, was a huge chocolate bar made by Hershey but covered in the omnipresent olive drab. It was 12 in long by 5 in wide and was about 1.25 in thick. This emergency ration bar contained vitamins and some sort of emulsion to keep it from melting. The emulsion made the bar so hard that even 20-year-old teeth couldn't dent it, so they had to hack off chunks, if necessary, with their knives and suck on the chunks. It tasted great though. Still farther down and on the sides of the pad were the cutouts that contained fishing hooks and lines, a compass, an extra box of .45-caliber ammunition, smaller scale paper escape maps, water purification tablets, quinine pills and Atabrine to combat malaria, a first-aid kit that contained two ampules of pain reliever morphine, matches, Zippo flints, a pack of condoms for use as water containers and other emergencies, signal mirrors, and miscellaneous other escape necessities.

As they approached the bridge at Kaifeng, Bill decided to take only his wingman, George, down with him to recon the bridge. He parked the other two up at 10,000 ft, just outside of automatic weapons range, and told them to circle and watch. As the two P-47s screamed toward the bridge at 340 or so mi/h, where they normally would have shot their guns to keep the antiaircraft gunners' heads down, Bill bet that if they didn't shoot, the enemy wouldn't either. The protecting Japanese troops were used to seeing bombers high up and would try to hit them, but they were not used to seeing a pair of fighters whiz by at low altitude. The ploy worked and not a shot was fired. A firsthand look convinced Bill that there was no way that even 12 fighters could do anything but take out a few spans of the bridge. Their 1000 pounders would bounce off the bridge's piers and abutments like snowflakes. Consequently, he and George climbed back to 10,000 ft to rejoin the flight and headed back to the west and home.

On the way over to Kaifeng, Bill had noticed an airfield symbol on his map about 10 mi west of the city, and he told his flight that they would check it out on the way home. So when they had finished looking over the bridge, he veered slightly northward and located a large grass strip airport. When they neared the airport, Bill saw a glint of sun off something and concluded that it might be an aircraft canopy. He immediately put his flight in trail and told them he might have an aircraft spotted on the ground. When he got closer, he spotted a Val Achai P-3A two-seater dive-bomber airplane near the edge of the airport. He instructed George to work over the hangar and the operations building to keep any gunners' heads down and instructed Roy and his wingman to hit the plane if they saw it or, if not, to shoot up the buildings or any other worthy targets.

None of the rest of the flight saw the Val on this pass. On the next pass, Bill brought them in from the east, because it was a major rule of survival never to attack an airfield from the same direction on consecutive strafing passes. Again, none of the others saw the Val, but Bill hosed it down again as he passed over. He saw it lean over and die. The other three were out of ammunition and he was flashing solid tracers, so it was time to head for home. Although it would not count as an aerial victory for ace status, it did count as an enemy aircraft destroyed. He was sorry that the other three guys had not been able to get a shot at it, but this was turning into a pretty good day. Nine locomotives and one airplane were not a shabby day's work.

After they passed over Anyang, then Loyang, and almost had the Yellow River bend in sight, with about 30 minutes left to home, they veered slightly southward to check the ferryboat crossing. This crossing was just barely within the Japanese lines and, being close to the lines, handled a significant amount of military vehicle traffic. All fighters passing that way had standing orders to check it out and attack if it had military cargo or if it was performing a hostile act. The definition of a hostile act included being in motion.

Sure enough, today there was a big, fat ferryboat just leaving the north bank of the river. The Whang Ho, which was running east and west just a few miles before turning northward for 300 or more miles, was about the size of the Mississippi near Memphis. Since no one else had any ammunition, Bill decided to park the other three guys at 8000 ft while he went down and shot up the boat. He throttled back to 20 in so he wouldn't get too fast in his dive. Above 320 or so, things happened so fast it was hard to get a big picture of what was going on and besides, if the air was choppy and bumpy, it was hard to held the pipper of the gun sight on the target. He lined the ferryboat up in his sights on the way down, and at around 600 ft, he began hosing it down. He fired short bursts until he ran out of ammunition at about 150 ft. It was standard attack doctrine to continue the dive in such a case, for pulling out right over your target left your belly exposed and your speed across the ground was temporarily very slow while you went through a 90° change in pitch angle. So Bill bore on down and leveled out just over the water with his prop just about to tick the water. His airspeed now read 360 mi/h.

Bill happened to glance up at the riverbank about 20 ft above him as he cruised by and saw a Japanese soldier tracking him with a rifle. He thought, "This guy looks as if he had shot skeet all his life! Hey! He's going to hit me!" Normally, infantry soldiers aimed their guns right at the place that they wanted to hit the plane, usually the pilot or his cockpit. Bill made some fast calculations. At 360 mi/h, he was doing 6×88 ft/s, which equaled 528 ft/s, and the Japanese soldier's gun would fire at 3000 ft/s muzzle velocity. Therefore, it would take the bullet 150/3000 seconds to travel the 150 ft separating them. In 0.05 seconds at 588 ft/s, the bullet should hit 24.6 ft behind the cockpit, or just past the rudder. Wrong. Not this soldier. He knew to continually lead the place that he wanted to hit by 25 to 30 ft. Bill was too low to even get his right wing up between his plane and the gunner, so he started to pull up slightly to give himself room to bank to the left. At this time, Bill heard a loud bell-like clang, felt a terrorizing thump on his back. and was aware that a cloud of dust was flying around in the cockpit.

Bill's first thought was of young Captain Harris, a West Point staff officer, who had been flying with their other squadron. Just last week, Harris had been hit in the upper leg near the hip and had been unable to figure a way to fly the plane and apply a tourniquet to the wound. The flight had hurried home as fast as they could go at low level, but they were 30 mi from home. Harris was losing so much blood that he chose to belly in on a sandbar in the Wei River. By the time the Chinese soldiers got to him, he had quietly bled to death while still sitting in the cockpit.

With this fresh in his mind, Bill came to the conclusion that the same thing would not happen to him if he were mortally wounded. He was going to jerk his beloved plane around, find the Japanese soldier, and cut him in two with his prop if he were in the open or, if he had run into the grove of nearby trees, he would root him out with his entire plane. As he hoisted the plane up and around toward the Japanese soldier, he felt for the hole in his back. With his flight gloves on, he couldn't feel a hole, but it did feel very sticky back there. He tore off his right glove and continued to explore the hurting area with his bare hand. It wasn't bloody and he couldn't feel a hole, so he concluded that it was not a life-threatening wound. Since he was not mortally wounded, he decided to let the soldier live. Bill called red flight and checked in. He told them he had been hit, but was apparently okay, and told them to rejoin him as he streaked for home.

As they bore on toward the base, Bill continued probing his back and became certain that the sticky stuff was sweat rather than blood. When he slowly started feeling better about his back, he realized that his right eye was smarting pretty badly. Apparently, it was filled with dust, sand, or something. He had always been a freak about caring for his eyes properly, and they were only 8 minutes from the base, so he resisted the temptation to rub the corner of his eye.

With the field in sight, they reported in to Sian tower and requested immediate landing to the west. They flashed in over the airfield at 20 ft, snatched the big fighters up into a tight left turn, lowered the gear at 1000 ft, then dropped the flaps and lowered the nose, and flew them smartly onto the ground. The guys watching on the ground as they peeled off and landed could tell that the flight had a pretty good day by the way they approached and attacked the ground.

By the time Bill's flight had taxied in and shut down the big Pratt and Whitneys, they had logged 7.33 hours and the crowd was gathering. Their squadron flight surgeon, Dr. Max Salvater, climbed up on Bill's wing and said, "Bill, let me see where you are hit." Bill stood up and started to peel down his wet flight suit so Doc Salvater could see his back. The Doc said, "What's the matter with your eye?" Bill replied, "Damn my eye! It's my back that I'm worried about, Doc." Doc Max looked at his back and said, "Your back has a bad bruise near the spine, but I don't see any holes. However, your eye is very bloodshot and I can see several fragments sticking in it. Get dressed and we'll get up to the clinic and have a look at it."

They walked to the clinic and Doc anesthetized the eye and examined it with a magnifying glass. He said, "Well, Bill, you finally got the Purple Heart I have been trying to give you for years." Bill replied, "You wouldn't dare put me in for a Purple Heart for a speck of dust in my eye, would you? We would both be the laughing stock of the entire 14th Air Force." Doc Max said, "Dust, hell. What I'm seeing here are seven jagged bits of copper and bronze metal, and if any one of them had been one eighth inch over to the center, they would have permanently destroyed your vision. Don't argue with me this time. I'm putting you in for a Purple Heart."

As the Doc continued to fish the fragments from Bill's eye, the head parachute rigger and his assistant came into the room looking like a pair of grinning possums. They said, "Captain Miller, here is what happened to your back." They held up the big Hershey bar, and in the center of the candy bar was embedded the steel core of a Japanese 7.62-mm rifle bullet. The core had penetrated about two thirds of the way through the bar. The four of them played detective that afternoon and came to these conclusions: (1) The first clang, or bell-like tone, Bill had heard was the bullet striking the edge of his armor plate, which was just behind the pilot's seat. (2) The thump and bruise on his back was the core of the bullet plowing into the blessed chocolate bar. (3) The fragments in his eye came from the jacket of the bullet, which had fragmented and ricocheted around the cockpit. (4) The Japanese soldier had been less than one quarter inch away from a confirmed kill because the hole in the right side of the P-47, the bruise on the armor plate, and the angle of the bullet all confirmed that the bullet was heading right for Bill's spine.

In later years, when Bill relived this mission, he often wished that there was some way to tell the lonely Japanese soldier how close he had come to being a dead hero.

Bad Weather

Told by Clifford J. Harrison, Jr.
First Lieutenant, U.S.A.A.F.

It was another early morning when three other pilots in D Flight of the 526th Fighter Squadron and I arrived on the flight line. I was still a relatively green pilot ready for about my fourth mission. We were outside our Quonset hut that we used for a briefing room, and I looked up at the sky. It was still dark, but I could tell that the weather was completely socked in. The clouds must have been about 1000 to 1500 ft. About this time, I had visions of getting back on the truck and returning to our squadron quarters and back to bed. Our flight leader, Captain Setencich, an excellent pilot and seasoned leader, said we should go ahead and take off, adding that we might find a hole and go up over the clouds.

We took off but could not find a hole. When it became obvious that we could not find one, he said we would go ahead and climb up through the stuff. That sent a few fear quivers through my body because I had never flown on actual instruments in weather. All the instrument flying I had ever done was either simulated in a Link trainer or under the hood in an AT-6 training plane with an instructor in the front seat. He told us to get in close. When he said that, I was very relieved because it meant he would be the one to fly on instruments and we would go through the clouds flying in formation on him. If I had not been so new on the job, I would have known this. He did not have to tell me to get in close more than once because I was not about to lose sight of him. I knew that if I did, two things would happen. First, I would have to fly on instruments myself, and second, I would be separated from the flight. Not liking either alternative, I put my wing practically on his lap.

We did climb through the layer of clouds, which seemed to me to take a much longer time than I am sure it actually did, and we finally came out on top. We then headed east toward the enemy lines. When we knew we were well into Germany and past the bomb line, which at that time was the Rhine River, we looked for a hole where we could go down. This time we found one, and our P-47s dove through it.

What we saw was amazing. At this stage of the war, the Germans tried to restrict their movements of vehicles to night when they were relatively safe from our fighter bombers. I believe that on that day I saw more movement on roads and rails than I saw on any of my other missions. We must have destroyed 20 trucks as well as worked over a number of trains. When we returned to our airstrip, we found that we were the only flight in the group that got a mission that morning.

Okinawa

Told by William H. Pickron
First Lieutenant, U.S.A.A.F.

In mid-1945, while based on a small island near Okinawa, I was directed to proceed to the island of Guam along with seven other pilots to pick up eight new replacement P-47Ns for our fighter group. As you probably know, in 1945 Guam was the home of General LeMay's famous B-29 bombers. Up to this time, being a fighter pilot, I had never gotten to know a bomber pilot, much less have a chat with one to find out what he did with all those throttles and engines. During the next two days, we were allowed to mingle freely with those famous crews. Some of the younger ones seemed a little vague about what the Navy and Army pilots had been doing in the Pacific prior to their arrival, but they assured us they would be able to take care of Japan post haste.

Guess you might say they were just about like any typical American man overseas in World War II. They bragged a lot, griped a lot, and said they really had it rough. We expressed our sympathy because they had to live in rooms in wooden buildings, some with electricity, fans, real mattresses, a sit-down "think tank" (toilet), with cold beer and colas on hand. They even had tablecloths and napkins, along with fried chicken our first night and baked ham the second.

Meanwhile, back at home base, our boys enjoyed sleeping on narrow hard cots under mosquito netting in hot tents, all of which were demolished in the "hurricane of '45." Most were frequently awakened in the middle of the night when "Bedtime Charley" dropped his nightly nuisance bomb. They got to stand in line with mess kit for Spam (baked, fried, broiled, or cold) or some other unidentifiable goo cooked outside in an iron kettle (possibly rations left over from World War I). Now do you get the picture?

After a short meeting of the minds, it was decided to relieve our B-29 friends of some of their misery and at the same time furnish a little happiness and cheer for our boys back at home base. About midnight, we borrowed a couple of Jeeps and liberated a substantial amount of beer from the rear storage room of the officers' club. On arrival at the flight line, the security guard let us pass after being assured that a final check and loading of all eight fighters must be completed at this time to make an 0430 takeoff. The beer was then placed in the wings of the fighters, after removing some 25,000 rounds of ammunition and leaving the same under the nose of a nearby B-29.

At 0500 hours the next day, our fighters joined on the wing of a B-29 navigation escort for the return flight home. Six hours later, our escort called, "I'll be leaving you now and returning to Guam. Your home base is ten o'clock 50 mi, broken clouds at 10,000 ft with haze." As we approached for landing, I could see a number of wildly maneuvering planes in the distance over Okinawa. On contact with the control tower, we were quickly asked, "What state?" This is asking for the remaining fuel and ammunition on board. My reply was "Two hours fuel and zero ammo." We were directed to land immediately. After landing, I returned to my tent and within 2 minutes the group commander's Jeep slid to a stop out front. His comments: "Lieutenant, I heard your conversation with the tower. Did you know Okinawa was under red alert? Furthermore, just what in the [censored] do you mean by flying into a combat zone without loaded guns? "Well sir," I replied, "it's been weeks since a Japanese plane has been seen in this area during the daytime and I thought our mechanics, who work day and night to keep us flying, deserved a treat. Therefore, we removed the ammo and loaded the wings with beer."

"Beer!" he shouted. "Yes, sir. We came home at 29,000 ft, and that beer should be about 2 degrees above freezing at this moment." He left in such a hurry I was covered with gravel from the spinning wheels of his Jeep, while his trail of dust led in a direct path to the flight line, where I feel certain he collected his share of cold beer.

I gather he wasn't too happy with me. However, to this day, I've never heard a word from him about my lack of ammunition on that flight. On the other hand, my feeling of accomplishment in delivering cold beer to the troops diminished somewhat 2 days later with news from a fellow pilot in a nearby fighter group. It seems the planes I had seen over Okinawa as I was landing were his flight of four engaging Japanese bombers. Each pilot downed two. Under normal circumstances, our flight very likely would have been part of that action were it not for too much beer and not enough ammunition!

A Walk Unspoiled

Told by O. T. Ridley
First Lieutenant, U.S.A.A.F.

The mid-April darkness seemed slow in coming. The voices on the road were becoming fewer and more distant. The forced rest period should have been appreciated, but I was anxious to be on my way. Earlier in the day, after I had bailed out of my P-51 and avoided the German search teams, I had found my way to this hiding spot. I was a short distance from a well-traveled unpaved road. It was interesting that there was no automotive traffic. I estimated my position to be about 20 mi west of Prague, Czechoslovakia. It was April 18, 1945. My Mae West, pistol, jacket, and brown flight suit had all been rolled into what I thought was an inconspicuous brown bundle. The bundle was similar to those carried by people on the road. The clothing I was wearing (that is, the pants and shirt) were made of wool and were olive drab in color.

When darkness fell and road foot traffic ceased, I stepped out on the road and headed west. Occasionally as I passed people, they spoke and I nodded as seemed customary. After an hour or so of walking, I came into the small town of Lany. It was about 2000 or 2100 hours and there were very few people about. The town seemed deserted. I walked through and into what I thought was the center of town and looked at the road signs. None of those names were on the nice silk map of Europe that came in my escape and evasion kit, which I still had. The maps I used for navigation had been lost in the parachute jump, along with my knife and other personal survival items. I proceeded west and took a turn up a street. Bad decision! It was a dead-end street that ended at a small pond. There was a two-story building on the right and a small dock-like structure that extended a short distance into the pond. There was one light on in the building to my right in the upper story. A small dog met me there. He sensed my apprehension and started to bark loudly. I spoke kindly and tried to move closer to him. The dog backed away and started to bark even more loudly! Another light came on in the second story. I turned and went back to the main road with haste and again headed west.

When well out of town, the last two men I passed spoke, and I nodded in return. They mentioned the word "Amerikanski" but made no effort to follow or stop me. I continued to walk in the opposite direction for about 100 yd. Still there was no move to follow. I decided that these men might be helpful. I turned and went back to them. I had found members of the Czech underground who had been out all day looking for downed airmen. One of these men was Mr. Jossef Kuna of Lany. The Czechs motioned that we should leave the road immediately. We moved to their left and into a slightly uphill open field. Just below the top of the rise, there was a deep ditch that separated an adjoining field. In this ditch, we were not visible from the road.

Previously, at our intelligence briefings at the squadron, we had been told that if we were shot down and contact was made with the underground, it was appropriate to give them your pistol, provided you felt secure in your position. I kept my .45 automatic with one round in the barrel. In the ditch, the Czechs talked briefly, which of course I did not understand. Then motions were made demonstrating possible injuries. I indicated I had no injuries. Then came signs and motions to indicate hunger. I had not eaten breakfast, and since this might be a long night, I indicated I was hungry. One man left to get the food and the other stayed with me. The man returned in an hour with the food, cold meat in a cup and a pint-sized metal container half full of what I thought was goat milk. I was relieved when my new friend returned with food and without Germans. I was not used to this type of food and was not very hungry. I forced it down, every bite and every swallow.

The Czechs let me know that we had to get out of the area. We went back to the road. The walking procedure was one man in front, my position 50 yd back, then the last man 50 yd behind me. If a car came by, I was to get in the ditch beside the road. Using this procedure, we walked about half the night. There were only two cars that passed during this walk. I got into the ditch on each occasion. Neither car slowed or showed the slightest interest in the walkers. We walked until past midnight and stopped halfway to the town of Ruda at a small, dimly lighted house.

Here we met two new people. These people had a dog that responded to uneasy strangers with barking that said, "I don't think that I will bite you, but you don't belong here." The two men that had walked me from outside Lany turned to go back after I was briefly introduced to the two new guides. Handshakes were made but no names were mentioned.

We used the same walking method, one man in front, one in back, with me in the middle. No cars passed on this leg of the journey. We came into a wooded area just before daylight and walked another 2 or 3 hours into deep woods. The interval between walkers decreased to the point we could keep the man ahead in sight. By midmorning, we arrived at a location in the woods where the Czech underground was hiding three escaped Russian soldiers. There was conversation between the Russians and the guides. The guides explained that I was a downed American pilot. This explanation seemed to make me acceptable. I settled in with the group.

The weather on my arrival was that of a nice spring day. However, I kept my flight suit and jacket in case of colder weather, which came later in the form of snow. The daily routine was to rise early and walk well away from camp into the woods. Usually, the Czech patriots would bring food during the day. I never knew exactly how they found us, or where they might be. We returned to our camp at dusk. The camp consisted of an underground cave. The cave entrance was a man-sized hole that opened vertically into a larger cave area. This area would accommodate four for sleeping purposes. There was a floor padding of straw, which added warmth. The area immediately outside the cave opening was given considerable care and attention so as to eliminate any indication of our presence. The last man to enter swept the entrance area of any footprints, broken twigs, and so forth. He then backed into the cave pulling the hole cover behind him. The hole cover consisted of sod and growing plants. It was about 2 ft in diameter and 13 in thick, and it matched the surrounding ground area. This duty was done by one selected for his special skills of arranging the cover to match the landscape and removing all traces of our presence. No one else was allowed to do this very important job.

I stayed with this group for a few days. The conversation was of the Russians coming in to take over Czechoslovakia, Germany, and France. This was my first experience of being with a group that did not speak my language. The general understanding of what was being said was much better than I expected. There was also the blessing of not being able to understand any matter that you did not want to understand.

When they found I was an officer, I was moved to another location about half a day's walk away but still in the same general area of the woods. The group here also consisted of Russians, but officers only. One man was a Russian submarine captain who had seen rough treatment at the hands of the Germans. He was childlike. The captain had problems finding his way about in the woods. Since I was new to this group, it fell my lot to take him into the woods each day with me. He was a tragic sight of a once proud and brave man. The other two men were Russian officers who seemed overly interested in the P-51, its cruising range, speed, and armament. This was particularly true of the self-appointed leader, who claimed to be a captain of artillery. When these types of questions arose, I did not understand the language.

We had the same underground sleeping arrangements as before. Food was brought in the same manner, but now twice a day. The food consisted of black bread, a butterlike spread, probably lard, cabbage, and sometimes potatoes. The artillery captain was not only interested in P-51s but seemed to be obsessed with sex. He demonstrated this with crude hand motions suggesting sexual intercourse. I did not understand. He certainly did not have any other peculiar traits. The other officer was able to help me understand that he wanted to know "how to make" an English-speaking woman. I let him know there were five English sentences that would do the trick. I held five fingers and tried to let him know these "pickup lines" had to be said properly and come from the heart. I pointed to my heart. (1) Oh, ugly one, your eyes remind me of the pools in the urinals in the Greyhound bus station. (2) Oh, ugly one, your aroma is that of 3-day-old dead fish. (3) Oh, ugly one, your face is that of a sick frog. (4) Oh, ugly one, your body is that of a time-worn old whore. (5) Oh, ugly one, if you were the last woman in the world, I would not screw you. We practiced these phrases at every opportunity. Finally, the English was very understandable and clear. I would hold up one finger for phrase one and the artillery captain would recite phrase one, two fingers for phrase two, and so on. Occasionally, I would stop to correct any flaw that might not enamor the lady in question. It passed the time. I have often wondered if he ever had the opportunity to practice his newly developed language skills.

Once, the underground people took us to a house for a bath. This was done at night with precautions; however, I was surprised that we went as a group. The tub was filled with water one time. They did not change the water, so we matched coins to determine bathing order. The benefits to the last man were questionable.

About 2 days later, I was moved to the home of Mr. Josef Pittner in the town of Ruda. The move was made at night with the usual precautions. One man in front, I was in the middle, and one man behind. The distance separating us was about 30 yd. Once again, no lights were used. There I was placed in the second-story upstairs attic, which was very comfortable compared to my past underground lodgings. In the left side of the attic, and well back under the eaves, was a large stack of hay. This haystack was my home for the next few weeks.

The Pittners brought food morning and night. Again black bread with a butterlike spread and cabbage or occasionally a potato. They made arrangements for me to get out of the attic at night to visit the outhouse. Sometimes I got to walk back and forth several times between the house and outhouse, a welcomed exercise. The Pittners had an English translation of the epic poem *Beowulf,* which I read again and again. It seemed like a good trade, my .45 automatic for *Beowulf.* The days passed pleasantly; early May was warm and sunny. The Pittners' home was a duplex and a German couple occupied the other side of the house. One warm day the children were playing outside in the yard, and the outside shutter door of the attic was banging loudly because of the wind. As I sat reading in the back part of the haystack, the woman came up the steps from downstairs to close the shutter door. She walked directly to the shutter, closed and latched it, turned to her right, and walked straight back to and down the attic steps. Had she turned to her left, she would have found me reading. On another occasion, the house was searched by the German military police. The Pittners had notification of their coming. I never knew how the Pittners got this information. On their arrival, I was well beneath the hay and far back toward the corner, directly under the lowest part of the eaves of the house. The whole haystack would have had to be moved to get to my position.

It was not unusual during early May to see German soldiers retreating through Ruda. The condition of the enlisted men was quite bad. However, the officers were well uniformed with boots shined, hair recently cut, clean shaven, and generally, they were riding well-groomed horses.

May 8, 1945, Prime Minister Winston Churchill made his speech declaring an end to the war. The Pittners listened on their radio and I was invited to sit in for the occasion. There was a large red "Verboten" sign pasted on the front of the radio. It was a warning, I suspect, of what one might or might not listen to. The speech was in English. After the speech, Mr. Pittner asked me if the war was over. I said, "Yes. The Germans had surrendered." Mr. Pittner decided it was proper for me to come down out of the attic. My time in the attic was not unpleasant, but confining. I did not give the matter of leaving the attic enough thought. Having left the attic, I went out into the street to look around and test my new freedom. Mr. Pittner quickly decided my departure from the attic might be premature. There were still German soldiers in the vicinity and the Russians had not arrived. I returned to the attic.

A few days later, the Russians came in and liberated the town of Ruda. I came out of the attic to stay. The German couple in the adjoining duplex seemed very surprised and fearful to find the Pittners had been keeping an American in the attic. I do not know how the German couple missed my earlier walk in the street. The Russians together with the Czechs, who had many scores to settle, hunted down the remaining German occupation forces. It was done much the same way we used to hunt rabbits in Tennessee. The Czechs and the Russians lined up almost line abreast and walked slowly through the woods and undergrowth. When they kicked up a German, if he ran he was shot and if he surrendered he was taken back to town.

There were a few innovative methods developed locally that made the hunt and the inevitable burial much more spirited. The burial detail consisted of a Russian and a member of the Czech underground, both well armed. They would escort two or three Germans out to the grave site. The Germans would then be instructed to dig the graves to exact specifications. The Germans, the likely grave occupants, were measured with the shovel handle for height and width. The number of graves dug did not always equal the number of Germans in the party. Occasionally, they dug one less. If the one to be shot was undecided, a coin was flipped to settle the matter. There was considerable talking and joking between the Russian and the Czech on the way to the grave site as to who should be shot and who should cover the graves with dirt after the shooting. Some Germans faced death bravely; others did not.

The countryside was generally peaceful now with the exception of an occasional celebrating Russian soldier. I gave my thanks and said good-bye to the kind and generous Pittner family. I then went back to Lany to visit with the Josef Kuna family. There I met Lieutenant Buford Stovall, a B-17 pilot who had been hidden in another small town nearby. Lieutenant Stovall and I visited in the Lany area for a few days, participated in the celebrations and parades, and visited the many friends of the Kuna family. Mr. Kuna knew of a truck that would be going to Prague in a day or so, and Lieutenant Stovall and I caught a ride on that truck. Once in Prague, we visited friends of the Kuna family and then reported to the Red Cross. Appropriate telegrams were sent back home. We traveled from Prague to Pilsen by truck. As we crossed the Russian-American line into Pilsen, it was indeed a pleasure to see a good solid American GI once again.

As I recall, I returned to military control June 8, 1945. From Pilsen I was flown with a group of POWs and evadees by C-47 to Camp Lucky Strike at Le Havre, France. After processing in Le Havre, I was allowed to go back to my fighter squadron, the 362d, at Leiston Air Base, Suffolk, England. Once there, I made a special effort to get reassigned to this fine organization, but I was not successful. I returned to Camp Lucky Strike and from there traveled by troopship to the United States.

In retrospect, I credit my successful evasion to a run of extremely good luck, superior training, and an esprit de corps instilled in the members of the 362d Squadron by men like Major John B. England, Major Leonard K. Carson, Captain John A. Kirla, Captain Robert D. Brown, Captain Charles E. Weaver, and finally and most importantly, to the heroic and unselfish patriotism of the people of the Czech republic.

Turkey Shoot No. 2

Told by William S. Miller
Captain, U.S.A.A.F.

One day several months after chocolate bar mission No. 62, Bill had his favorite flight out on a deep penetration into the eastern part of China. On that interesting mission, while around 600 mi from home base, they had discovered a rail line that started southeast at Shijiazhuang and wandered in the direction toward Jinan and Jining. No one in their squadron had ever worked over this rail line, so they felt that it might be very productive. Besides, they would be within 170 mi of the coastline and they might find some elusive Japanese fighters to play with.

They were loaded with a 75-gal drop tank on each wing, which would give them a good 7 hours of flying time, unless they had to use full power and water injection for a significant time. If that were to happen, they could always climb up to 20,000 ft or so and lean the birds out to 60 or 65 gal/h.

This spur line of around 160 mi did not show up on their navigational charts, but this was not unusual. Frequently, towns were shown on the wrong side of rivers, and the charts had many large white splotches which indicated this territory was unexplored. The country they would fly through would gradually go from hills and small mountains into low marshy land. This was part of China where the Yellow River is famous for its flooding.

They cruised at 12,000 ft today instead of the normal 6000 to 8000 to save fuel. On the way to Shijiazhuang, they spotted two trains and naturally they went down and hosed the locomotives but didn't waste much time getting back to 12,000. At Shijiazhuang, they descended to 4000 so they could see targets better. Sure enough, about 10 minutes after they started down the track, they found a 13-car passenger train. It was heading north and Bill told the second element to shoot the locomotive while it was on a long, low bridge.

George Frieze and Ritchie did a good job as usual, and the engine toppled over into the marshy water. Bill then set up another ground gunnery training command traffic pattern and they began to work over the passenger cars. Unfortunately, this time, there were not ravines or surrounding hills to hold the Japanese into the area. The swampy water appeared to be less than 18 in deep and this slowed them down a little. But it did not prevent them from spreading like wildfire.

On Bill's sixth firing pass, targets were becoming scarce and they were generally just splashing up water and mud. The people were learning to get down into the water to escape their .50 cals. As he looked very hard on the next run, Bill saw one standing target. It was a woman with a baby in her arms. This is probably the smartest thing she could have done and it probably saved her life. Bill told his flight to start looking carefully because there were many ladies and children among the survivors.

Ritchie and Emreigh both replied, "But as you have told us many times, Chinese civilians have been warned to stay off trains, trucks, planes, or anything with an engine." Bill responded, "Yes, but I can't tell if these are Japanese or Chinese civilians. Anyway, I am not going to shoot a lady with a baby." Ritchie and Emreigh were both getting good and they were starting to feel pretty cocky, so one of them said, "To hell with that noise."

With that, Bill rocked his wings and said, "Join up into a spread formation. We are heading home by way of Kaifeng and Louyang." On the way home, they found three more locomotives to destroy, but this did not placate the hungry wingmen. The troops, except for George, kept grumbling, and finally one of them said, "Whatever happened to our Wild Bill Miller? Is he turning into a pussycat?"

Bill let the guys have their grumbling with good humor, for he sort of felt that their criticism was deserved. On the other hand, he was glad they had quit shooting. They discussed how many Japanese they had found on the way home. The consensus was that there were at least 125 people in each car for a total of over 1500. They estimated that they killed at least 60 percent of them and wounded another 25 to 30 percent.

Bill thought long and hard about the words of his upset wingmen on the way home and said to himself, "They may be right. I only need three more missions to 75 and I may be getting cautious and losing some of the edge." He had been thinking of getting a transfer to one of the long-range fighter units in Iwo Jima or one of the other Pacific islands, and he decided to get the ball rolling by discussing it with Flight Surgeon Max Salvater when he got home. He did so want to have a go at some Japanese fighters before this war ended, and he was convinced he would never find one in China.

Flattop Landing

Told by William T. Sellers
Lieutenant Junior Grade, U.S.N.R.

The latter part of July 1945, operating off the coast of Japan, the fleet would strike for 2 or 3 days and then withdraw to refuel, rearm, resupply, and so on. During one of these refueling operations, I was assigned to go to a small carrier and fly back to the *Independence* with a replacement Hellcat.

This story won't mean a thing really to anybody unless they have been a carrier pilot. In the first place, until you experienced a sling ride from the aircraft carrier hangar deck to a destroyer, you have missed out on one of the great episodes of life. You may make it safely, you may get your feet in the water, or you may be dragged like bait on the trolling line, but that's another story.

Anyway, I left the *Independence*, safely transferred to the destroyer, and then transferred from that destroyer to a flattop to pick up a new F6F5 Hellcat. After signing papers and making sure the plane had a plotting board secured in the cockpit, I was catapulted from the flattop. After becoming airborne and contacting the *Independence*, I was directed to what was commonly called the "lost sheep circle." This is a designated area on a certain bearing from the carrier where you circled until you were directed to land. When the fleet is refueling, the ships move at a slow speed and all turn together on command, so it followed that I might have to spend some time in the air before the carrier would be ready to take me aboard.

After I received the radio signal, I had "Prep-Charlie," which means you leave your position, come to the carrier, and fly a landing pattern in relation to the carrier as if you were landing. The landing pattern for the carriers is an elongated oval. You come by the bridge of the carrier about 150 ft, going in the direction the carrier is headed, and fly 1 or 2 minutes in that direction, make a left-hand 180° turn, come back until you get approximately even to the bridge of the carrier, and then make another 180° turn to your left so you come in behind the carrier to land. You make this pattern even though the carrier is not ready to take you aboard. I noticed the ship was kicking up more wake, so I knew it was beginning to pick up speed, and I got a signal that I had "Charlie." This means the carrier was being prepared to take me aboard.

I was 2 minutes ahead of the carrier starting my left-hand turn when I could see the ship begin its turn into the wind. I adjusted my height, speed, and position in relation to where it would be when it had completed its turn into the wind. The carrier picked up speed so it would have about 30 kn over the deck when it completed its turn into the wind. Of course, it left the protection of the other ships and became a little more vulnerable to any submarine that may have been shadowing the fleet.

Seeing the carrier start to turn into the wind, I had been able to determine the direction of its course from the foam running off the tops of the waves. I followed my adjustments and got all the things ready to make a carrier landing, such as tail hook extended, landing gear down, cowl flaps open, and oil flaps open, so hopefully I'd end up in the proper position to come aboard. I was still receiving signals indicating "fouled deck," which means the ship is not ready to take you aboard, all the way up to my groove where I was supposed to come aboard when I got a clear deck signal.

I picked up the landing signal officer with his paddles and he gave me the cut. Down I went and caught the wire. Following the usual procedures, as the tail hook engages, you lean forward and open the cowl flaps, which you closed at the last minute, unlock your wings, let the plane roll back while the deck man disengages your tail hook, bring your tail hook up, and then follow the deckhand's motion forward until he signals you to stop and cut your engine. I really didn't think too much about it at the moment of landing, but I did not know that the ship had already begun to slow down and turn back into its correct pattern. As I was getting out of the plane, low and behold, here comes the executive officer off the bridge and down to the flight deck to give me a "well done and congratulations." As the carrier had been into the wind a total of fewer than 15 seconds, he said that it was the quickest he had ever known of a replacement plane coming aboard and the shortest time for a carrier to be out of position.

With all the computers we have today, I like to recall how all the calculations, both knowingly and unknowingly, were made for me to get that plane into landing position just as the carrier hit the wind line. I still think a human brain is the greatest calculator there will ever be.

Commander Bud Schumann

Told by David C. Kipp
Lieutenant Junior Grade, U.S.N.R.

I would like to tell you a story about one of the bravest and calmest officers I ever had the pleasure to serve under. Commander Bud Schumannn was our fighter squadron commanding officer. On July 14, 1945, while flying from the carrier U.S.S. *Hancock* (CV-19) off the coast of Japan, a special flight took off at 0800. It was to be a fighter sweep of eight F6Fs. Commander Schumannn was the flight leader. He was the only one in the flight flying the rather new F6F5E with the radar dome on the right wing.

We flew toward Hokkaido, the northernmost Japanese island, about 100 mi from the ship. We had good weather with some rain and clouds. We strafed trains and ferryboats full of Japanese soldiers as well as planes parked on some Japanese airfields. Some of the planes on the ground were destroyed and would not even burn when we blew them up because the Japanese were so short of fuel. There were no fighters in the air to hinder us, so we did a lot of damage.

After the strike, we all joined up in formation and started back to the fleet and our carrier. The weather took a turn for the worse. It was raining and we had to climb higher to stay in the clear. We donned oxygen masks and climbed to 25,000 ft. We heard from our ship that the weather there was very bad (100-ft ceiling and half-mile visibility).

Here we were, somewhere over the fleet at almost 30,000 ft, low on fuel, with no way to find our ship. At this crucial time, our skipper came over the radio, clear and calm, with all the reassurance of a father talking to his sons, and said, "This is what we will do. I have the carrier on my radar screen. They are into the wind and ready to take us aboard. I want all of you to get behind me in single file. I will let down through the storm and as I approach the ship I want the last plane in the string to land. I will see that the last plane clears the water by 100 ft. We will make eight passes at the carrier. If you don't have enough fuel to go around and make another pass, wiggle your wings on final approach and they will not give you a wave-off." He was so calm and sure of himself that we almost believed him.

We started down into the driving rain and lightning. All that each of us could see was the trail of lights of the plane above us. Down, down we went. Finally, he called for the lowest plane to leave us and land, which it did. I was about in the middle of the formation, so I only had to go around three or four times. Finally, my turn came. Bud Schumannn came on the radio and told me to leave the formation, that I should be able to see the carrier. Sure enough. There she was directly ahead of me in the rain. I made a perfect approach because my red fuel warning light was on and I knew I didn't have enough fuel for another approach.

I landed safely, parked my plane, got out, and went to see if my buddies would get down okay. The weather was so bad I couldn't see the top of the bridge on the carrier. They did all get aboard safely, in order, and as the last plane, Bud Schumannn's came aboard and its engine shut down, a huge shout of joy went out from all of us. That one man saved all eight of us that terrible day. None of us will ever forget him.

How to Land a PT-17

Told by Dutch McMillin
Captain, U.S.A.A.F.

Graduating from flying school in 42-F, I was assigned to the 89th Fighter Squadron of the 80th Fighter Group at Mitchell Field, New York. After a reasonable training period, we were transferred to Republic Field on Long Island. All of our training in New York was in P-47 Thunderbolts.

About that time, Madame Chiang Kai-shek visited President Roosevelt with an urgent plea for more help in the Far East. The 80th Fighter Group was sent to Richmond, Virginia to check out in P-40s and head for India. The three squadrons and group headquarters were soon settled in Assam. The 88th Squadron was near Shingbriyang, the 89th at Nagaghuli, and the 90th at Jorhat.

We had a Stearman PT-17 among our airplane stock. It was often used to ferry ground officers (paddlefeet) from here to there around the territory. One day, it fell my turn to take an armament officer from our base at Nagaghuli to Jorhat, about 100 mi. The trouble was that Jorhat was closed by the weather to any P-40 flying, in fact to all flying. On the other hand, our base at Nagaghuli was wide open. This sent Lieutenant Paddlefoot and I on our way. We found the airstrip all right by using the iron beam (railroad tracks). The strip was socked in with a 100-ft ceiling. Once committed, I continued to try to put her down. Now the PT-17 was a notorious ground looper. I put her down in a heavy drift. She took over and ground looped us all. From going in one direction, we were suddenly going 180° in the opposite direction. It turned out to be a controlled ground loop. The wing didn't drag the ground and no damage was done.

We taxied up the strip to the point where every grounded pilot of the 90th watched the proceedings with glee. I taxied up, cut the engine, and climbed out. Before any of my colleagues could say a word, I said to all in a loud voice, "I just wanted to show you b_____s how short you can land one of these things."

The Last Plane
Shot Down

Told by William T. Sellers
Lieutenant Junior Grade, U.S.N.R.

Air Group 27 was aboard the carrier *Princeton* when it was sunk in the South Pacific. The air group was returned to the states to be re-formed, and I joined it in Maine in January 1945. I served for several months at Patuxent Naval Air Test Center in Maryland when I was assigned to the air group.

Following 2 months in Maine, the air group was transferred to Creeds Field in Virginia. I won't say anything about Creeds Field other than the runway was nearly as wide as it was long. In May, we went to Hawaii and down to Hilo for a short period of time before we got orders to return immediately to Pearl Harbor, where we boarded C-54s and island-hopped directly to the Philippines.

Air Group 27 went aboard the carrier *Independence*. The fleet left the Philippines the latter part of June and conducted continuous operations off the coast of Japan until the surrender in September 1945. On August 10, 1945, three divisions of four planes each, flying F6F Hellcats from the *Independence*, were conducting fighter sweeps over the central part of Japan. At this time, I was either flying wing on the division leader or leading the second section, I don't recall which on this particular flight. In a strafing run on an airfield, I became separated on the recovery from the rest of the flight. I spotted another F6F pilot from the *Independence*. We joined up together and headed back toward the fleet. Before we reached the coastline, we picked up a straggling torpedo bomber and decided we would escort him back because he would make an easy target by himself.

Sometime after we passed the coastline headed toward the picket destroyer, a plane passed right under us with big red circles painted on each wing. It was a Zero. My fellow fighter pilot and I immediately left the torpedo plane to fend for himself, and we did a quick 180 to chase the Zero. The other fighter plane was a little faster, and he was beginning to get ahead of me so I dropped my belly tank and eased past him. He realized what I had done, and he dropped his belly tank and eased past me again. About this time, we both went to water injection in order to catch the Zero before it got back to the coastline.

His plane got behind the Zero first, fired a burst, and got a little smoke, but the Zero did not change direction. Then the other fighter broke off his attack and I slid in behind the Zero and fired my .50-caliber machine guns. The Zero began to smoke heavily. I can remember screaming, "Burn, damn you. Burn." I was expecting to see a fireball at any instant, but although he flamed brilliantly, he did not explode. The Zero pulled up into a left wingover, the canopy came off, and it headed straight into the water. I did not see any parachute from the plane. My fellow fighter and I rejoined and proceeded to the destroyer. In due course, we landed on the *Independence*.

I asked the plane captain to be sure to save the gun camera film, as it had a Zero in it. This was on August 10, 1945, which was the same day they dropped the second atomic bomb. They had dropped the first atomic bomb on my birthday, August 6. The gun camera film turned out to be of good quality, and we showed it in the ready room several times. I brought it home. However, it has now been lost and I have regretted that I did not have a blowup of one frame showing the flaming Zero.

The war ended shortly thereafter, so I believe, although I am not sure, that this Zero was the last plane shot down by a carrier pilot in World War II. Although Air Group 27 had many kills in its previous tour of duty, this was the only plane shot down by this group on this last tour. Saw one plane, got one plane: That equals 100 percent.

The Atomic Bomb

Told by Eugene D. Pargh
Lieutenant Junior Grade, U.S.N.R.

It was April 1, 1945. Our squadron, VC-13 on the carrier *Anzio*, joined in the initial air strike on Okinawa. We had four fighters and four torpedo planes and were sent in to attack two Japanese ships pulling out of the harbor. I was flying one of the Grumman Wildcats. My squadron skipper, Lieutenant Bob Williams, an old man of 25, was leading us across the southern tip of Okinawa. Antiaircraft fire opened up on us pretty heavily, my skipper made a 90° turn, and we took an extra 10 minutes going around the island.

I was the second pilot to go in on the ships. On my first pass, I was so intent on getting a hit with my rockets that I went in too low and pulled out almost too late. We were told to pull out at 1000 ft, but I ended up at 50 ft. On my second run, I was out at 1000 ft. I still remember the tracers going by, but we were all lucky. No one was hit and we left both ships sinking.

Every day for the next 6 days, we hit everything that moved on Okinawa. Twice I landed on other carriers, once because of a deck crash on the *Anzio* and once because I was almost out of gas. The tough part of each day at Okinawa was getting back to our carrier. Our battleships and cruisers were shelling Okinawa and shooting at every plane that came near them, and we had to fly between them and Okinawa. We quickly learned to stay away from our own ships. They were shooting at everything in the air. We always moved closer to Okinawa, as their fire was much less than that of our own ships.

I'll never forget April 6, 1945. That day, I shot down three Japanese planes. Two of our destroyers, about 25 mi north of Okinawa, were under attack by enemy planes. I picked up one after he strafed one of our ships, and after a few bursts from my .50-caliber guns, it was in the sea. I pulled up and found myself being fired on by our own ships. With good reason, the boys on the destroyers were trigger happy, and they too shot at everything. I pulled away from the ships, and as I was climbing, I got another Japanese plane in my sights and quickly got No. 2. Again, pulling up and away, I saw another enemy plane heading for our destroyers. I got off a couple of bursts, but had to pull away as the destroyer was firing at both of us. I remember the "friendly" tracers going by and knew I had to get out quick. A few minutes later, I gained some altitude, got another Japanese plane in my sights, and quickly got No. 3. A short time later, I got a short burst off on another enemy but soon was out of ammunition. Another Wildcat came in and shot this one down.

I was lucky enough to join up with one of the Wildcats from my squadron, and we sweated our way home. We were north of Okinawa and again had to head south between our battleships and Okinawa, being shot at from both sides. We finally found the *Anzio*, landed okay, and headed for the ready room. It was just another normal day. You were either hit or not. No big deal.

We were with the fleet in Japanese waters the day we received orders to head for Guam, where a replacement squadron would take our place aboard the *Anzio*, and we would fly home. That day, the ready room was full of talk about what planes we would be coming back in when we were sent back. At that time, it looked as if it would take a couple of more years, as the Japanese were not about to give up. An eventual invasion would cost at least a half million American and several million Japanese lives.

On the day after we left for Guam, there was an announcement on the ship's loudspeaker that we had dropped an atomic bomb on Japan, and we had been ordered to rejoin the fleet. They wanted to keep all ships at sea. One of the boys in my squadron came up to me and asked, "What is an atomic bomb?" I told him I didn't know. No one on our ship knew. We were all upset about not going home right away, but it ended all talk about what plane we were coming back in next time. At Okinawa, we lost only two planes. One was shot down by one of our night fighters, and the other by one of our ships.

Prior to being in the Pacific, I had duty aboard a baby carrier, the *Guadalcanal,* in the North Atlantic on antisub patrol. There was no tougher flying duty in World War II. Just making it back to our carrier and finally getting aboard safely was a daily challenge. We were caught in a hurricane, and 2 days later, we were almost to Iceland before we could turn the ship and head south. Baby carriers were not built for that kind of weather. After finally making it back, even Norfolk looked good. After the North Atlantic, being in the Pacific was a piece of cake.

On antisub patrols, a torpedo plane and a fighter would go on 100-mi searches. The torpedo planes had a pilot, a gunner, and a radar operator and were a safety net for the fighters. So naturally, we "hugged" our TBMs. If we lost them, getting home was more of a problem. Our Grumman Wildcats were great. While slower than the Hellcats or Corsairs, they could outmaneuver anything we had, could take a beating, and survive a rough landing, which most were in the North Atlantic.

The Wildcat did have one small problem. My automobile has twice as many instruments as a Wildcat. We had an airspeed indicator, altimeter, radio, revolutions per minute indicator, a turn and bank indicator, and a radio direction finder that worked by Morse code. The code was changed daily and showed us the direction to get back to our carrier. That was great, but it went out if the radio went out. This happened a few times and was a constant fear. In the North Atlantic, we would not last 10 minutes in the frigid water if we went down. When our engine would cough, which happened often, we stopped breathing. But my Wildcat always got me home. Others in my squadron were not so lucky.

The training we received and the lessons we learned at Georgia Preflight School and through graduation at Corpus Christi always stayed with me. I learned never to give up, never to quit, and always to do my best. The lessons I learned were with me in the Navy and are still with me more than 50 years later.

In the service, none of my many close calls bothered me. I guess that is why only young kids make good fighter pilots. However, a few years later, I started having nightmares. These dreams continued for about 5 years.

The 3 years and 4 months I spent in the Navy will always be with me. Those who were not with us would never understand. I remember the boys I served with and the things we did better than I recall what I did or whom I met last week. It was the greatest 40 months of my life, but I would not want to do it again.

Over the years, I have heard and read of many well-meaning people talk about how awful it was for us to drop the atom bomb on Japan. Obviously, those people were living safe lives during that period. It was Truman's greatest decision. It ended the war quickly and saved millions of lives.

Many years later, in 1987, my wife and I took a cruise from Hong Kong to several stops in China, with a 1-day stop at Nagasaki in Japan. Our tour bus took us to the Memorial Park at the site where the A-bomb hit Nagasaki. A Japanese lady gave a speech describing the bomb blast and the horrible damage it did. There were all kinds of exhibits showing the terrible destruction caused by the bomb. I wanted to tell her about the Navy pilot who crash landed at sea near one of the islands held by the Japanese. When our troops invaded a week later, he was found with his head chopped off. I'm sure there were thousands of stories like that. As we wandered through the park, one old-timer about my age asked me what I thought about all this. I said it was justified. He said he felt the same way, but stronger. Very few people in the present generation of young people would ever know or understand how we feel.

Pucker Factor 9

Told by William H. Pickron
First Lieutenant, U.S.A.A.F.

An interesting thing happened to me on my first wedding anniversary, August 12, 1945. I was a member of a WWII fighter group based on the small Pacific island of Ie Shima, located near the northern end of Okinawa. Our arrival, as it turned out, was just in time for the last 2 months of the war. We flew the AAFP-47N on missions over Shanghai, Korea, and Kyushu, the southern part of Japan.

On this particular day, our targets were a petroleum refinery and a large aboveground tank farm located in the northern part of Kyushu. On arrival, each fighter dropped two napalm bombs and made several passes firing a series of rockets and many rounds of .50-caliber ammo. The explosions and raging fires resulted in a successful first part of the flight and cost the Japanese dearly in gasoline.

The second portion of the mission consisted of strafing any and all targets for some 200 mi on the way out. Returning in flights of four, we had planned our departure to include a fast run across an airfield complex near the city of Kagoshima, located on the southwest end of Kyushu. This site was apparently used by the Japanese as a staging base for night raids on Okinawa, and we had hopes of finding a few planes on the ground. There were no aircraft to be seen, so we strafed hangars, ground vehicles, and what seemed like hundreds of antiaircraft guns of all sizes, each determined to make our visit permanent. This will not be news to lots of fighter pilots, but in my view, the tracers—those little white things floating past the cockpit and in front of my enlarged eyeballs—seemed so thick that I could almost get out and walk on them.

On leaving the airfield, we flew down a valley with high mountains on each side, a route that would take us to the ocean and a direct path to home base. Alas, this was not to be. Within a short distance, we encountered clouds blocking our flight from the mountaintops to the ground. Reversing our direction required a reduction in airspeed, which resulted in recrossing the same airfield at a much slower speed than desired. We fired our remaining ammo, and I believe they must have used most of theirs on us.

You might say that we entered the unplanned and unscheduled portion of the mission at this point. On clearing the field boundary, we were met head-on by a flight of Japanese fighters. The total ammo remaining among our flight was expended when one of my guns fired about 10 rounds. Then it really was quiet on our western front. As they pulled up, we dived for the deck with throttles bent full forward. This is where the "pucker factor," on a scale of 1 to 10, hovered around 9 for what seemed like an eternity as we were fired on from the rear. You might associate pucker factor with the word *fear*, in this case accompanied with the silent question of "How did I get in this mess?" We must have inherited the luck of the Irish, or they were very poor shots, because we finally outran them and make a safe return home. It was an interesting wedding anniversary, and I doubt if there will be another quite as exciting for me.

Halsey's Whimsy

Told by David C. Kipp
Lieutenant Junior Grade, U.S.N.R.

It was September 1, 1945. On the aircraft carrier U.S.S.
Hancock, the Navy pilots were all feeling relieved and happy
that the war in the Pacific was over. They were looking forward to celebrating and dreaming of going home to their
families. I was one of the F6F fighter pilots in Air Group Six.
Over the bullhorn that evening came the announcement that
all pilots were to report to their ready rooms at 1900 hours.

While there, we were told that General Douglas MacArthur
and all of the senior officers of the fleet would be on the battleship *Missouri* off the coast of Japan to accept the surrender
of the Pacific war by the Japanese empire tomorrow. All pilots
would take part in a massive flyover above the battleship. This
was to be our last flight! And believe me, none of us were very
excited about it!

The next day, we were up at dawn and every possible plane
on the carrier was made ready for flight. The weather was definitely not in our favor. Light rain was falling, visibility was
almost zero, but off we went! Joining up in our formations
was almost impossible. Because of the clouds and rain, we
could hardly see each other. But we had to show the Japanese
our mighty power, so we continued on our mission. We flew
for 4 hours over Tokyo Bay and over the battleship *Missouri*
while MacArthur accepted the unconditional surrender from
the Japanese emperor.

This was the greatest assembly of carrier air power in the history of aviation. Some 1200 aircraft of 15 air groups took part in this mass flyover above Tokyo Bay. The air was thick with planes! Luckily, I got back to the carrier safely with most of my fellow pilots. But after the war was over, we heard there were as many as eight to ten collisions, and a lot of the pilots were killed on the morning of September 2, 1945.

Twelve hundred pilots called this last flight "Halsey's Whimsy." Is there any other possible term that could have been used?

Epilogue

I feel privileged to have been asked to compile and edit these interesting reminiscences from World War II, as written by the dedicated flyers themselves. I thank my friends for having given me this honor.

In the Introduction to this collection of true stories, we included a small portion of poetry foretelling what was to come many years later. In like manner, two selected verses from "Recessional" by Rudyard Kipling provide an appropriate conclusion to this anthology:

The tumult and the shouting dies—
The Captains and the Kings depart—
Still stands Thine ancient sacrifice,
An humble and a contrite heart.
Lord God of Hosts, be with us yet,
Lest we forget—lest we forget!

Far-called, our navies melt away—
On dune and headland sinks the fire—
Lo, all our pomp of yesterday
Is one with Nineveh and Tyre!
Judge of the Nations, spare us yet,
Lest we forget—lest we forget!

John K. Breast

The Men of the Middle Tennessee WW II Fighter Pilots Association

THE MIDDLE TENNESSEE WWII FIGHTER PILOTS ASSOCIATION, established in 1992, is an eclectic fraternity of doctors, bank presidents, architects, attorneys, engineers, professors, and, naturally, pilots living within the Nashville area. Initiation into their elite ranks occurred fifty years earlier, when each member flew fighter aircraft in combat during World War II. Collectively, they flew nearly every type of fighter in every theatre of war, and won almost every medal that can be awarded a pilot. What bonds them today is the one thing they share: electrifying firsthand accounts of aerial combat.

ROLAND HENRY BAKER, JR.: Ensign Baker was born October 10, 1923, in Newton, Massachusetts. He joined the Navy on January 1, 1943, and served until May 1, 1963. He was awarded the Distinguished Flying Cross and Air Medals while serving as a carrier fighter pilot. He served in World War II and the Korean War. After retirement from the Navy, he became chief pilot for the Marston Corporation, captain for Cape and Island Airways, and chief pilot for New England Propeller. In 1975, he moved to Nashville, Tennessee, to become chief pilot for Northern Telecom. Tragically, Roland Baker was killed in a motorcycle accident on June 12, 1994. He is survived by three grown children and his wife, Kay Baker.

E. H. BAYERS: Commander Bayers enlisted in the Navy in 1928. He attended Navy Flight School and took part in the Battle of Midway. He led the Fighting Three Fighter Squadron. He retired with the rank of captain.

GEORGE M. BLACKBURN: Major Blackburn was born in Nashville, Tennessee, on March 31, 1922. He flew 90 missions in a P-47 Thunderbolt with the 66th Squadron, 57th Fighter Group in Italy. He earned the Distinguished Flying Cross, Air Medal with five clusters, Presidential Unit Citation with cluster, the Unit French Croix de Guerre with palm, and European Combat Ribbon with three battle stars. He has a Bachelor of Arts and a Master of Arts from Vanderbilt University and lives in Brentwood, Tennessee.

ROY B. BROSTER, JR.: After the war, Captain Broster returned to Nashville, Tennessee. He was employed by the Third National Bank where he held the position of vice president in the commercial and branch division at the time of his retirement in September 1980. He and his wife, Mary Ann, live on Tim's Ford Lake in Franklin County. Their son, John, is an archaeologist with the state of Tennessee.

BILL BURCH: Commander Burch entered the Navy in July 1941 and received his wings in July 1942. He flew Grumman F4F Wildcats in combat in the Pacific and shot down three Japanese Zeros in one flight.

JOHNNIE BRUCE CORBITT: Captain Corbitt was born on November 7, 1922, at Plant, Tennessee, a small farming community in Humphreys County. He attended public schools in Humphreys County. He enrolled in the Civilian Conservation Corps in July 1940, and later worked briefly for the Tennessee Valley Authority. He was accepted into the Air Corps as a cadet on July 4, 1942. His first assignment following flight training was as a target tow pilot at Matagorda, Texas. He served with the 493d Fighter Squadron of the 48th Fighter Group, Ninth Air Force. After the war, he attended Austin Peay State College and received a Masters Degree from George Peabody College in 1950. He was recalled to active duty as an aircraft controller during the Korean War. He was returned to reserve status in 1953 and he was employed by the TVA before returning to teaching high school in Madison, Tennessee. He retired from the Air Force Reserve in 1967 with the rank of Lieutenant Colonel and retired from teaching in 1983. He is presently "happily retired from everything."

FRANK N. DAVIS, JR.: Frank Davis joined the Navy and completed his pilot training in 1943. He served two tours of duty in the Pacific. After the war, he joined a reserve squadron in Memphis and was recalled to active duty during the Korean War. He remained in the squadron and flew jet fighters until 1957. He was employed by Bell Telephone and retired in 1984.

HERMAN K. FREEMAN: Lieutenant Colonel Freeman joined the Air Corps in March 1942. He flew in the European theater of operations for 16 months and completed 149 combat missions in the P-40 and P-47. He returned to active duty in 1952 and remained on active duty until his retirement in 1970. He then worked for several charter airlines until his final retirement in 1991.

LEE V. GOSSICK: Captain Gossick served as a fighter pilot in World War II flying P-40s in North Africa and Sicily. He retired from the Air Force in 1973 with the rank of Major General. Following World War II, his assignments included duty as commander of the Arnold Engineering Development Center, director of the F-111 program, and commander of the Aeronautical Systems Division at Dayton, Ohio. Following his retirement from the Air Force, he served as assistant director of regulations, U.S. Atomic Energy Commission. He retired from federal service in 1980 and joined Sverdrop Technology, Inc. He served as vice president and deputy general manager of the AEDC Group and retired from that position in 1988. He resides with his wife, Ruth, in Tullahoma, Tennessee.

ROBERT F. HAHN: Hahn returned to the United States in September 1945, and was assigned to the Air Force Training Command. He instructed in Stearmans, T-6s, P-51s, T-28s, and T-33s. He also served in Korea in 1953 and retired from the Air Force in 1964.

CLIFFORD J. HARRISON, JR.: First Lieutenant Cliff Harrison is a native of Nashville, Tennessee. He is a graduate of Vanderbilt and the Nashville School of Law. He spent his entire working career at Third National Bank in Nashville. He retired at the end of 1987 as vice chairman. Since his retirement, he has become an avid photographer and has had several exhibitions of his work.

EDWARD F. JONES: Captain Edward Jones flew 76 missions as a P-47 pilot. He attended Washington and Lee University in Lexington, Virginia. He was employed as a newspaper reporter by the *Nashville Banner* and served as public relations officer for the Tennessee Department of Safety. In 1958, he was named director of a congressional traffic safety committee and served 2 years in Washington, D.C. He also served as executive vice president of the Nashville area Chamber of Commerce. In 1987, he was named editor of the *Nashville Banner*. He is married to the former Wanda Smith Tindell.

DAVID C. KIPP, SR.: After the war, Lieutenant Junior Grade Kipp was sent to Pensacola, Florida, to instruct. He remained in the Navy until June 1947. He then returned home to Michigan and went to college on the GI Bill at the University of Michigan. On December 17, 1949, he married Viola Hazelaar. He served in the Naval Reserve at Grosse Isle, Michigan, and was called back into the Navy in September of 1952 for the Korean War. In 1957, he was sent to Pensacola for 8 weeks to learn to fly helicopters. By then, he had moved to Nashville, Tennessee. For 6 years, once a month, he drove from Nashville to Detroit to meet with his helo squadron for 2 days. He was promoted to commander and made skipper of HV-731 for 3 years. After that, he remained in the reserve and retired with 27 years active duty and reserve. In Nashville, he joined a real estate firm, Dobson and Johnson, and sold residential real estate for 35 years. He retired in 1995 as a senior vice president. He and his wife will celebrate their 50th anniversary December 17, 1999.

BEVERLY W. LANDSTREET: Captain Landstreet served as a Marine combat pilot in the South Pacific theater. He retired with the rank of captain, U.S.M.C.R. He attended the University of North Carolina and graduated from the Nashville School of Law with a Bachelor of Law degree. He serves as a director of Kentucky Home Mutual Life Insurance Company and Kentucky Home Capital Company.

J. Pat Maxwell: Major Maxwell was born in Pleasant Hill, Alabama. He graduated from Auburn University. After the war, he served as a mortgage loan officer with The National Life and Accident Insurance Company of Nashville, Tennessee. He presently lives in Nashville, Tennessee, with his wife, Ruth.

Elbert "Dutch" McMillin: Captain Dutch McMillin graduated from Duke University in 1942. He joined Les Brown's orchestra briefly before the outbreak of the war. Returning to civilian life in 1945, he resumed his career as a saxophonist and clarinet player for WSM's Walking Crew. In 1956, he began a long and successful career as head of the New England Life Insurance Agency. He passed away in 1995 at age 79.

William S. Miller: Captain Bill Miller graduated from the Army Aviation Cadet program in March 1943. He flew in Panama until being assigned to the 81st Fighter Group in Italy. He flew 40 combat missions in the P-39 before being assigned to China where he flew 35 more. He attended Middle Tennessee State University and graduated from Vanderbilt. He also received a Master of Business Administration degree from George Washington University. He returned to active duty in 1950 and served for 24 additional years, retiring as a colonel. He accumulated over 6000 flying hours in 51 different aircraft models.

Charles R. Mott, Jr.: Captain Mott served as a fighter pilot with the 58th Fighter Squadron of the 33d Fighter Group. This was one of the few units to participate in combat action against all of the Axis powers. Captain Mott was awarded the Distinguished Flying Cross, the Air Medal with two oak leaf clusters, the Presidential Unit Citation with oak leaf cluster, and six battle stars.

Eugene D. Pargh: Lieutenant Pargh was born in Cleveland, Oklahoma, in 1922. He moved to Nashville with his family when he was 17 and has called it home ever since. Prior to enlisting, he worked for Martin Aircraft in Baltimore and Vultee Aircraft in Nashville. In 1946, he began building homes and later developed residential subdivisions. He and his wife, Madeline, have three children and seven grandchildren.

Ernest C. Perry: After the war, Captain Perry went into residential and commercial land development and building in Nashville, Tennessee, and Huntsville, Alabama. He continued to fly as a pilot and squadron commander with the Tennessee Air National Guard. He is still active as a pilot and aircraft owner. He has three children and four grandchildren. He is currently managing his commercial real estate interests with his son. Captain Perry would like to recognize his brother, Coleman, who was awarded two Purple Hearts while serving with the Seventh Army in France. He feels that men like Coleman made a much greater contribution to the war effort than he did.

WILLIAM PICKRON, JR.: First Lieutenant William Pickron received his military pilot wings in April 1942 at the age of 19. He was the youngest pilot in the Air Corps at that time. He served as a fighter pilot from 1943 through 1945. He retired as a lieutenant colonel after 26 years of service. He continued flying for 12 more years as the chief pilot of the state of Tennessee.

WILLIAM A. POTTS: After World War II, William Potts married Lorena Ray whom he had known since childhood in his hometown of Dickson, Tennessee. They have two daughters, Ann and Carol, and now four grandchildren. After being in and out of several businesses, he organized and built AM and FM radio stations in Dickson, operated other stations, and had interests in others. After retiring, he enjoys golf, photography, travel, and raising and eating hot peppers. As he said at the close of a radio program, "Thanks for listening. Don't forget to grin. So long everybody."

O. T. RIDLEY, JR.: First Lieutenant "Tom" Ridley retired from the Tennessee Air National Guard in February 1977 as commander of the 118th Tactical Airlift Wing with the rank of brigadier general. He served in World War II, Korea, and flew troops and cargo into Vietnam with the Tennessee Air Guard during that war. Presently, he is adjunct faculty member at Middle Tennessee State University with the Aerospace Department. He enjoys the outdoors and continues to canoe, hike, backpack, fly fish, and shoot (long-range target). He has a current rating of Certified Flight Instructor, Glider, which allows him to keep a hand in aviation and participate in an interesting sport. He is a lifetime member of the Air Force Association and the 357th Fighter Group Association, a member of the Retired Officers Association, the National Rifle Association, and president of the Stones River Chapter, Sons of the American Revolution in Murfreesboro, Tennessee. He is married, has three children, and two grandchildren.

WILLIAM T. SELLERS: Lieutenant Sellers entered the University of Virginia after the war. He graduated with a law degree in 1950. He has practiced law in Murfreesboro, Tennessee, since that time. He has four children and seven grandchildren.

WILLIAM SHWAB: Lieutenant Shwab went into residential building and land development in Nashville, Tennessee, and Huntsville, Alabama, following the war. He retired in 1972 but continued to fly as a hobby until 1992. He has three children, two grandchildren, and one great-grandchild.

ROY D. SIMMONS, JR.: Lieutenant Colonel Simmons was born in Bowling Green, Kentucky, in 1922. He graduated from flight school on August 30, 1943. He joined the 111th Tactical Reconnaissance Squadron in Pomigliano, Italy, in February 1944. He completed 110 combat missions. He retired from the Air Force on August 1, 1967. Following his retirement, he became the dean of students at Massosoit Community College, Brockton, Massachusetts. He retired from this position on February 1, 1981, and moved to Nashville, Tennessee. He and his wife, Glendell (Jennie), live in Nashville, Tennessee.

ENOCH B. STEPHENSON: After the war, Major Stephenson entered the investment business. He continued to fly with the Tennessee Air National Guard as a squadron, group, and wing commander. He retired in 1972 with the rank of brigadier general. He has two sons and two grandchildren.

JOE THOMPSON, JR.: Major Thompson received a Bachelor of Arts degree from Vanderbilt University in 1941. He served in the European theater of war earning six Bronze Stars, the Distinguished Flying Cross, the Air Medal with 15 clusters, and the Croix de Guerre.

LESLIE E. TRAUGHBER: After the war, Captain Traughber attended Vanderbilt undergraduate school, received his Medical Degree at the University of Tennessee in Memphis, and interned at Jackson Memorial Hospital, Miami, Florida. After 4 years of general practice in Aiken, South Carolina, he completed a residency in anesthesia at the University of Georgia. He then returned to Nashville in 1958 to practice at both St. Thomas Hospital and Donelson Hospital. He retired in 1987 at age 65. He and his wife, Pat, have four daughters and seven grandchildren who reside in Nashville.

WILLIAM W. WELLS: Captain Bill Wells was born in Nashville, Tennessee, in 1923. After leaving the Army, he graduated from Vanderbilt and went to work for Genesco. He was recalled to active duty in 1950 and served as a jet fighter pilot in Korea. He married Joan Welby in 1952, and they presently live in Franklin, Tennessee.

KENNETH WEST: Lieutenant Junior Grade Ken West is a Kansas native who joined the Navy in 1942. He was assigned to Fighter Squadron VF-15 in September 1943, with a Pacific combat tour in 1944. A Kansas State University graduate with a Bachelor of Science in architecture, he worked as an architect until retiring. He and his wife, Carolyn, live in Nashville, Tennessee.

W. S. WHITMORE: Lieutenant Bill Whitmore joined the Navy cadet program in 1942. He was assigned to fly Corsairs patrolling the many "bypassed" islands during the U.S. offensive. At the conclusion of hostilities, he taught school until being recalled to active duty during the Korean War. He was employed by the Federal Aviation Administration in 1961 and retired in 1992.

HENSLEY WILLIAMS: Major Hensley Williams was born on April 18, 1917, at the Traveller's Rest in Davidson City (Nashville), Tennessee. He attended the Vanderbilt University School of Engineering, graduated from Annapolis in February 1941, and received a Master's Degree from George Washington University. He was assigned to the South Pacific in 1942 flying F4U Corsairs. He also served in Korea and Vietnam, retiring after 34 years of service. After retirement, he worked with the Tennessee Conservation Department for 8 years.